GREEN
&GOLD

John Travers is an international energy expert from Dublin. A chartered engineer and graduate of UCD, he began his career at Shell International. He worked on oil and gas fields in the Northern Netherlands and subsequently led work on commercialising novel technology in The Hague. He left Shell to do an MBA at Harvard and write his first book, *Driving the Tiger, Irish Enterprise Spirit*. He then joined McKinsey & Company, management consultants, to lead projects as diverse as energy, not-for-profit healthcare and satellite design in Europe, the US and Africa. The founding CEO of Alternative Energy Resources Limited (AER), he has grown the company successfully since 2006.

GREEN & GOLD

Ireland a Clean Energy World Leader?

John Travers

The Collins Press

FIRST PUBLISHED IN 2010 BY
The Collins Press
West Link Park
Doughcloyne
Wilton
Cork

British Library Cataloguing in Publication Data

Travers, John J.
  Green & gold : Ireland a clean energy world leader?.
  1. Renewable energy sources—Ireland. 2. Power resources—
  Ireland. 3. Energy policy—Ireland.
  I. Title
  333.7'94'09415-dc22

ISBN–13: 9781848890435

Typesetting by Carrigboy Typesetting Services
Typeset in Berkeley Book
Printed in Great Britain by J F Print Ltd.

Contents

For John and Mary

Acknowledgements

Thanks to John, Mary, Sharon, Ken, Penny, Gavin, Fiona, Niall and Rosalind for their support and also for reviewing the book's content. Special thanks to Karen for her encouragement. Thanks to Kerry O'Sullivan for detailed review of the data and to Pearse Buckley, Amanda Barriscale and the team at the Sustainable Energy Authority of Ireland for comments and feedback on some sections. Thanks to the whole team at The Collins Press. Thanks to Kieran, Ronan, Eimear, John, Patricia, Richard, Audrey, Daragh, Paul, Anne-Margaret, Owen and Denise for their support and help.

Thank you to all those who helped me with support, information and advice, including: Kjell Aleklett, Gerard Boylan, Tom Bruton, Sharon, Ken and Melissa Byrne, Colin Campbell, Aidan Corbett, Niall Crowley, Karen Devine, Eamonn and Niamh Devine, Niall Devine, Simon Dick, Laura Dillon, Daragh Downes, Steve, Cathy and Isla Ellis, Andre Fernon, Shay Garvey, Stephen Hall, John Heffernan, David Horgan, Mark Hughes, David Hynes, Eamon Lalor,

Owen Lewis, Richard and Jennifer Linder, Penny, Gavin, Jason and Emily Marrow, Conor McCague and the team at Business and Leadership for use of the photo, Ronan, Eimear and Carla McArdle, Kieran McDermott, John, Patricia and Cillian McManus, Mary Meaney, Dipti Mehta, Jim Mountjoy, Clare Mulvany, all the team at NovaUCD, Andrew Parish, John Roberts, Audrey Ryan, Paul, Anne-Margaret, Sara, Sophie and Oliver Saunders, Eugene Smith, Brendan Sullivan, David Sullivan, Jack Teeling, John Teeling, Conor Toolan, Fiona Travers, John and Mary Travers, Rosalind Travers, Maria Tuohy and Kartik Varma.

Introduction

'The challenge, in short, may be our salvation.'
 – John F. Kennedy, addressing the US
 National Academy of Sciences, 1963.

When future generations recall our period of life on earth, what name will they give to our era? Heavy footprints have been laid on our recent path by world wars, space travel, quantum physics, information technology and genetic engineering. However, these imprints are faint compared to the impact that fossil fuel energy has had on our contemporary history. In time, our era will be known as The Carbon Age. It is an age that is ending.

Look around you. Almost every material aspect of life in the developed world is shaped by the energy latent in fossilised carbon or its derivative products. The influence is so prevalent that we take it entirely for granted. The electricity in a light bulb and every electrical component, the machine power or heat that moulded almost every object around us, the central heating or air conditioning, the production of food, most plastic items, the fuel that propels the vehicles and planes outside: life as we know it has a source of creation in carbon.

The energy and materials for the life we know have been sourced primarily from hydrocarbons such as oil and gas. These have provided the fuel for the fastest economic and social growth in the history of humankind. Total world gross domestic product (GDP) per capita grew at some 0.6 per cent on average per year in the century prior to the mainstream adoption of oil and gas, around 1900. During the subsequent century, the average rate of growth tripled to 1.9 per cent per year. It is no coincidence that the average annual world population growth rates over the same periods were 0.5 per cent and 1.4 per cent respectively. During this carbon age, the total population has grown an unprecedented fourfold, from some 1.7 billion to nearly 7 billion.

Carbon energy has not only driven this development but has also brought many benefits. It has brought the world closer through air travel; given us the freedom of the open road; provided warm and comfortable homes on well-lit streets; energy for industry; communications networks and functioning medical equipment; and materials for pharmaceutical, agricultural and personal luxury businesses. However, carbon energy has also brought a terrible legacy.

The burning of carbon to release energy in recent decades has been so intense that the addition of carbon dioxide to the atmosphere has exceeded its natural absorption, like a bath filling faster than it can empty. The amount of carbon dioxide in the atmosphere is now greater than at any point in measured history. If temperature increases resulting from the greenhouse effect of carbon dioxide pollution continue to rise at current rates, respected scientists tell us that we can expect vast tracts of land to be covered by rising seas, major droughts, food and water shortages, displacement of large populations and international conflict over scarce resources. It is not a comfortable picture. Are the scientists right? The evidence is compelling.

Furthermore, this carbon-based energy source is now running out. No one disputes that the finite reserve of fossil fuels will dry up; only the date is in question. Experts debate a window of thirty to fifty years, even when unconventional sources such as tar sands and oil shale are taken into account. The demand on these falling reserves continues to rise as a result of explosive population growth and consumption beyond reasonable needs in the developed world.

Fossil fuel's two fatal flaws – dwindling reserves and a negative impact on the natural world – precipitate the end of the carbon age and present the single greatest challenge to human development. We have no choice. If we wish to continue to develop society and live without chaos in nature, we must find and use alternatives to carbon energy.

What is energy? Where does the energy we use come from? What are the possible alternatives for heat, power and transport to which we must turn? Will renewable energy meet our needs and how might the nuclear option work? This book addresses these questions. It first provides a clear picture of the source and use of our current energy. It explores our voracious thirst for the hydrocarbons that fuel our economy and lifestyles. Ultimately, it assesses, with optimism, the practical energy alternatives and efficiency savings that can be applied to meet all our needs.

The book will show how energy is both Ireland's greatest challenge and greatest opportunity. It lays bare unprecedented energy and climate change crises that may yet ravage the nation. It describes that while Ireland has become heavily dependent on volatile foreign fossil fuel, importing almost 90 per cent of its energy needs, it has over three times accessible renewable energy more than total requirements on its own doorstep. Ireland is endowed with some of the most powerful wind and waves on the planet, some of the highest yields of next-generation sustainable biomass in the world and plentiful solar energy, despite the clouds. The accessible renewable energy is a small subset of the full renewable resource, which can be easily captured, without a wind turbine blitz, without competing with food and without prohibitive costs. The wealth of Ireland's accessible renewable energy is similar in scale to production from the massive nation-shaping oil and gas fields of the North Sea or the Middle East – but instead of polluting and dwindling, it is clean and perpetual. This book shows how 20 per cent of Irish energy needs can be met by renewable energy by 2020 and sets a vision for meeting 80 per cent of energy needs by 2050. It highlights that employment can be provided for more than 80,000 in the near term and an enormous fifth of GDP can be derived from clean energy exports.

Irish green energy offers a golden opportunity. Ireland can choose to overcome the energy challenge it faces and in achieving this, can become an outstanding world leader and global beacon of future energy.

ONE

Energy is . . .

'. . . like the God of the creation . . . within or behind or beyond or above his handiwork, invisible, refined out of existence, indifferent . . .'
    – James Joyce, *A Portrait of the Artist as a Young Man.*

Energy is everything we know. It is within or behind everything that is created and changes. Energy is the driver of all motion. It is the life of all biological things and innate to all matter. It is light and every living sense. It is the flickering neurochemical reactions that create thought and consciousness.

The ancient elements of earth, water, air and fire are all direct manifestations of energy, each in a constant state of flux, changing from one form to another. Energy is the surge of that flux through time, like the current of a river into which we can never step twice.

Does energy help define or shed light on the unsolved mysteries of creation? These are the things that we do not know. They belong to the realm of Kierkegaard's infinite resignation and to faith. Energy is simply everything we know.

WHAT DO WE KNOW? . . . ENERGY SCIENCE

From a scientific point of view, energy is defined as the ability to do work. One unit of energy, the joule (J), is exerted when a unit of force moves an object one metre. Energy exerted over time is measured in watts (W). One joule of energy exerted for one second is one watt, while one watt operating for an hour is called a watt-hour (Wh). In the world we know, we describe energy in many familiar forms. For instance, *heat energy* is measured as the temperature of things and is in fact a measure of how quickly atoms are moving or vibrating in a substance. *Gravitational energy* is the attraction of masses to each other across a distance, such as the pull between an apple on a tree branch and the earth or the moon and the earth's oceans. *Electrical energy* is established by the flow of electrons along a conductor. *Nuclear energy* is released by splitting or fusing the nuclei of atoms and the spontaneous decay of radioactive atomic particles.

More generally, energy can be classified as either *potential energy*, when something is at rest but has the potential to do work, or as *kinetic energy*, when something is in motion and work is being done.

A book held above the ground in a pair of hands has potential energy. Although it is resting, the book has the potential to fall closer to the earth if the hands were to open. Energy would be exerted if the book were to fall to the ground. Until it falls, that potential energy is stored, waiting for a trigger to convert it to another form of energy.

Kinetic energy is experienced, or 'work is done', when the book falls. Roughly one joule of work is done if a small book falls one metre. When the book clatters to the floor, the kinetic energy of motion ceases and the energy is converted to a little heat from friction, some sound energy and a new source of potential or stored energy in the changed profile of the atoms of bent or torn pages.

CONVERTING ENERGY TO CHANGE LIFE

The science of energy becomes life changing when we harness energy as it is being converted from its potential to its kinetic state to do useful work on our behalf. Potential energy resides like a locked-in tension in the chemical bonds of molecules of all matter. If matter is converted from one form to a simpler form, it may release energy in the process. For example, when wood is combusted with oxygen, the carbon in the wood combines with oxygen to form carbon dioxide, water and ash residue. However, the stored energy in the chemical bonds of the carbon dioxide, water and ash output can be shown to have less potential energy than that stored in the original wood. The difference in potential energy between the start and the end of the process is released primarily as heat and can be put to good use.

3

In the same way, when other forms of hydrocarbon, such as petrol or coal, are combusted with oxygen, heat is again released from the potential energy in the chemical bonds. This heat can in turn be converted to kinetic energy. The heated expansion of air can push a piston in a car engine or the boiling of water can create steam to drive a generator for electricity.

Converting energy from one form to another is never fully efficient. Some energy is inevitably lost and may be partially converted to a form that is not useful. In the case of the car internal combustion engine, not all of the heat energy is converted to kinetic energy in the moving pistons. Some is lost as heat to the surroundings. In fact, a typical car internal combustion engine loses 75 per cent of the energy potential of its fuel and it is never captured again.

The mainstream fuels of the modern world are those substances that can easily and cheaply be converted and that release useful energy in the process.

CONSERVATION, DISORDER AND THE LAST ACT OF TIME

Energy conversion complies with a key principle. The total amount of energy does not change but is conserved. This constancy is true of the total energy in the universe. If one part of the universe experiences an increase in energy, another part must lose the same amount of energy. In science, this is known as the first law of thermodynamics. In everyday life, it means that energy cannot be created or destroyed but simply changes form.

The Latin poet Ovid suggested that nature turned all chaos to order at the time of creation, hurling the rough unordered mass of things in the universe into their rightful places and binding them in harmony. For millennia, philosophers found comfort in this same hope – that order follows chaos, that the future will be safe. Sadly, the opposite metamorphosis is true of energy and all life.

The physicists of the nineteenth century showed that all things naturally progress towards total and utter disorder. All energy and matter dissipate to the state of greatest chaos. Heat energy creeps from warm to cold bodies, never the other way around. Everything disintegrates. In science, this is the second law of thermodynamics. In life, this is the inevitability of complete anarchy. This disorder, or entropy, is driven by energy changing from one form to another, from high intensity energy to lower intensity, distributed energies. Natural disorder is also what we often call beauty, such as the crumbling russet of scattered leaves or the tumultuous rush of falling rivers.

Humans have generally endeavoured to place order on things. We have fashioned objects, cities and civilisations from natural resources, harnessing energy to help us do the work. The irony is that we have been wading against the overwhelming universal flow of increasing disorder: our harnessing of energy conversion actually accelerates the process.

The conversion of energy is an advanced definition of time itself. Time moves forward as energy conversion drives the increase of disorder in the universe. Measurement of time as the revolution and orbit of planets marking days and years or a pendulum swinging between states of potential and kinetic energy striking out seconds and minutes has been displaced: time is now measured by the precise frequency at which electrons change energy levels in an atomic clock.

The conversion of high intensity to low intensity energy to do work is a one-way process. Those sources on which we have readily laid our hands are degraded once converted and are less useful for doing work. In the same way, all energy conversion in the universe will lead to a point of least usefulness and maximum disorder at some point in the future. Science tells us that once maximum entropy is reached, absolutely everything will have a uniform heat, density and even colour, which, astronomers say, will be sandy beige. In the end, the last act of energy will be the last act of time.

$E = mc^2$

When we think about the kinetic energy of the falling book, we can intuitively understand that the energy in its motion should relate to the mass of the book and the speed at which it is falling. The bigger the book, the bigger the crash. The less speed it has managed to work up, the less the impact. It can be shown by classical mathematics that kinetic energy is equal to the mass of the moving body multiplied by the square of the velocity, divided by two (i.e., $KE = \frac{1}{2} mv^2$). This confirms our insight that the bigger an object and the faster it is moving, the greater the energy. Think of the energy in an impact of a fast-moving juggernaut hitting a lamp post compared to a slow-moving bicycle hitting a lamp post. The equation does not yet convey any equivalence between energy and mass.

Einstein formulated a theory that had a revolutionary impact on this definition of energy. In his spare time while working in a Swiss patent office, he set his mind to understanding why the speed of light appeared constant in all circumstances, even when one is moving relative to the direction that light is moving. Intuition suggests that the relative speed of light might well increase while moving towards it and slow down while moving away from it. This happens to the relative speed of sound, for instance when we are moving closer to or further away from an ambulance siren. Major experiments that had expected to find changes in the speed of light by measuring it under different frames of reference failed to detect any change, such as an experiment conducted by Albert Michelson and Edward Morley in 1887. No physics puzzle had ever generated more intense effort to explain why an expected result did not happen. Albert Einstein entered the scene in 1905, at the age of twenty-six, to provide an answer.

While others considered how best to measure a changing speed of light, Einstein considered what it would mean for space and time if the speed of light were indeed constant. In his special theory of relativity, he suggested that the only way to understand the experiment results was to say that the laws of physics are the same in all inertial frames of reference and in particular, that the speed of light is the same regardless of how one is moving relative to it.

This completely shattered the understanding of our world held for hundreds of years and still challenges some of our very basic instincts when considering it. For instance, it suggests that time does not progress steadily but actually depends on the speed at which one is travelling.

For instance, take the case of one person sitting on a train moving towards a source of light while another person sits on a platform. Despite the fact that one person is moving towards the light and one is stationary, the speed of light is unexpectedly the same for both. Speed is determined by the distance travelled and the time taken to travel it (i.e., speed = distance / time). If the speed of light is the same for both the stationary and moving observer but the distance is changing between them, then time must also be changing between the observers in order for the above equation to remain true. Time is actually moving slower for the person on the platform compared to the person on the train. The phenomenon becomes noticeable at speeds approaching that of light.

When energy was analysed under this new theory of relativity, it became clear that the classical equation of $KE = \frac{1}{2} mv^2$ did not hold up entirely. At speeds approaching light, the accuracy of the equation faltered. Einstein worked to find a more correct definition for kinetic energy under his new theory. He found that the kinetic energy of a moving object was proportional to the speed of light. Furthermore, he showed that an object's energy at rest was equivalent to its mass multiplied by the speed of light squared, or mc^2. This groundbreaking development showed that mass is actually a possible manifestation of energy.

Since light travels at the enormous speed of some 300 million metres per second, the energy of even a small amount of mass is also enormous. This is fundamental to nuclear energy, the source of the sun's energy and man-made power and destruction. Einstein continued to develop his relativity theory and created the mathematical basis for predicting the Big Bang theory.

GENESIS OF ENERGY

The Big Bang was the explosion of birth when everything that we know in the universe was created. This phenomenal explosion was accompanied by overwhelming heat energy and instability. The entire contents of the universe expanded away from the source of the explosion and began to cool. Atomic particles such as quarks and gluons were formed within seconds. Within three minutes, nuclei of protons and neutrons were being formed by superheated nuclear fusion. This soupy broil continued to swell and larger masses congealed out of the fog of particles and elements. After a billion years, the first stars were born and seeded the first galaxies. Thirteen billion years later, in the universe of today, galaxies continue to hurtle away from each other. They are accelerating and are driven by an energy form recently detected, called 'dark energy'. Though little is known about it and its secrets are still being uncovered, dark energy is known to comprise a staggering 73 per cent of the entire universe.

Echoes of energy unleashed at the Big Bang churn in the star closest to earth. The sun was formed 4.5 billion years ago from a swirling cloud of hydrogen gases and particles. This matter exerted a gravitational energy that packed it closer together and increased its temperature. The heat rose to a point where nuclear fusion of hydrogen began and resulted in a temperature explosion that counteracted the continuous pressure of gravitational collapse, achieving relative stability. There is enough nuclear feedstock to maintain this intense balance of gravity and thermonuclear energies, and sustain our sun for another several billion years. The enormous well of nuclear energy produces heat energy at temperatures of 5,500 °C at the sun's surface and up to 15 million °C at its core. The molten pummelling of elements sends searing heat and light energy 150 million km through space to earth and light years beyond. The energy that reaches our planet is the genesis of our energy and life.

PLANT ENERGY

Light energy from the sun is captured by plants and converted into chemical energy by the process of photo-synthesis. Light enters green-coloured chlorophyll that is part of the chloroplast in plants. The chloroplast is like an 'energy factory', where water absorbed from the environment is split apart by the light energy. The chemical energy and electrons that previously bonded the water are ultimately gathered up in energy carriers, called NADPH (nicotinamide adenine dinucleotide phosphate) and ATP (adenosine triphosphate).

Meanwhile, carbon dioxide and more water are absorbed from the environment. The carbon from the carbon dioxide is affixed to existing organic compounds and the energy carriers impart their energy to these compounds to produce simple sugar, life's building block, as well as oxygen. The simple sugars and oxygen produced have greater net chemical energy than the starting products. Sunlight supplies the required energy difference.

The sugars created by photosynthesis provide all the starting material needed to build complex carbohydrates and all other parts of the plant. With sufficient water, carbon dioxide and sunlight, plants flourish and evolve in profuse variety. The first green plants of algae are thought to have evolved 1 billion years ago. The first land-based, mossy plants appeared some 475 million years ago. There have been countless cycles of life, decay and rejuvenation. Plant carbohydrates laid down in thick layers over these millions of cycles and fossilised are the primary finite hydrocarbon fuel sources of the modern world.

Plants, acting as solar energy catchers, are indispensable for animal life. Animals cannot convert light energy into chemical energy directly themselves so they collect their energy by eating it as food. The oxygen given off by light energy conversion in plants throughout the world is the same oxygen animals require for respiration.

BIOLOGICAL ENERGY

The body is, in a functional sense, a mobile energy converter. It is constantly converting the chemical energy of food to do useful work such as moving muscles, activating nerves in the body and brain, maintaining body temperature or building new cells. The full set of chemical energy conversions occurring in the body is called metabolism, which is derived from the Greek word *metabole*, meaning change or conversion.

The pathway of energy conversion starts with breaking down food while chewing. In the stomach, it is further reduced by acids and enzymes that literally cleave their way through large molecules. Finally, the basic components of food are liberated as simple sugars, fats and proteins. These are the building blocks for the biology of the body.

Simple sugar, glucose, combusts with oxygen that is inhaled into the lungs, absorbed into the bloodstream and carried to where the glucose fuel-stock is waiting. The combustion, or respiration, results in the creation of water and carbon dioxide that have a lower stored chemical energy than the starting glucose and oxygen. This energy conversion is the reverse of photosynthesis, which synthesises sugar and oxygen from water, carbon dioxide and sunlight.

The energy released from food is put to work to sustain life. The fuel conversion is similar to petrol combusting with oxygen to release heat that moves a car engine piston. However, in the body's case, energy released in controlled combustion can be gathered up as a reserve in the energy carrier, ATP (adenosine triphosphate). The energy carriers store and deliver energy as it is needed. They can couple with proteins to trigger muscles, transport substances through membranes or mend and grow cells.

When a body is exercising vigorously, the breathing rate increases to feed oxygen to the furnace of energy production. If the body uses energy faster than it can be available, energy reserves are depleted and functions become limited and lethargic. When exertion is stopped, gasping for oxygen continues to enable the restocking of ATP. If available glucose for energy production is used up, the body will turn to fat as a source of energy.

The average energy intake required by a 70 kg person to do a normal day's activities is about 2,400 kcal. This is about the same energy rate as a 100 W light bulb. Running 1 km will burn some 75 kcal, roughly the same energy in a slice of bread. Burning 1 kg of fat (or 2.2 lb) releases 9,300 kcal. Therefore, to lose 1 kg of fat, one would have to run 1 km every day for 120 days or eat one less slice of bread each day over the same period. Happily, there is another option. It has been calculated that 75 kcal is roughly the amount of energy released by ten minutes of laughter.

While energy in the body is converted primarily to do work, the energy that is transformed to less useful heat must be dissipated. Some heat is exhaled or carried away by ambient breezes. If needed, the body can boost heat transfer by opening pores to release warm perspiration. When less useful energy such as heat is taken into account, the overall efficiency of the human body in converting food energy into work is some 25 per cent, about the same as a petrol car engine. How much of that work is truly useful or wasted when applied is entirely subjective.

TAKING IT ALL FOR GRANTED

The processes of plant photosynthesis and living cell respiration act in perfect symbiosis. Each creates ingredients for the other to occur. Photosynthesis requires carbon dioxide and water and releases sugars and oxygen. Respiration requires sugars and oxygen and releases carbon dioxide and water. Sunlight is the energy source that feeds the virtuous cycle. It is a delicate harmony that we simply overlook or ignore as we carry on our everyday lives. Aside from the intricate energy balance of our bodies and nature, we generally do not bother to think about man-moulded energy sources and uses. When we turn on the kettle, fire up the computer or turn the ignition in a car, we do not think about the energy behind these acts. It is just there so we just use it. There is some vague association with energy and conservation, especially when a bill arrives, but our thoughts on energy are usually as fleeting as the flick of a light switch. We simply live the lives and lifestyles to which we have become accustomed, engrossed in the everyday effort of hope and survival. We have enough to be getting on with each day. Energy is taken for granted.

However, the impact that energy has on our daily lives and the world around us is profound. From the biochemical energy that sustains our every moment to the heat, power and transport of society, to nations that rise and fall over control of fuel, energy is the factor more than anything else that shapes us.

WHAT DO WE KNOW? . . . ENERGY HISTORY

Every meaningful technical invention ever made in the history of mankind has related to a new approach to harness energy or increase energy efficiency. Managing energy is a core facet of human nature.

The first energy we knew was food. We gathered and then hunted it to sustain life, build families and start communities. A captured spark of fire provided the energy to heat food and travel further into wintry lands with its protective warmth. We fashioned tools to reduce the energy burden of daily survival. Humans gathered by choice to share collective energy for a greater good. The energy of oxen, horses and other animals was harnessed to carry loads and work the land, with the first image of a man riding a horse at Susa, an ancient settlement in modern day Iran, dated to 4000 BC. The management of energy gave people more time to devote to leisure, science, philosophy and arts from which culture and civilisation grew. Energy was also the currency of power. Slavery was first inflicted by people who valued the control of human energy more than the freedom of life. The latent energy of the catapult and the tightened bow decided wars and established empires.

Waves of progress accompanied the invention of energy-saving devices such as the lever and the wheel. They enabled movement of goods, trade and epic construction projects such as the Irish passage tomb at Newgrange from 3200 BC, and later the Egyptian pyramids from 2600 BC and the palaces of Assyria (Iraq) from 2400 BC. The leading engineer of the Roman Empire, Vitruvius, said in 27 BC that without contemporary energy-saving machines, 'We should not have plentiful food … enjoyment … transport … and without them every kind of work is difficult'.

A shift from applying the animate energy of muscle power to exploiting the inanimate energy of natural resources set humans on a track of accelerated growth. Where water fell freely, its energy was directed to the grinding of grain in watermills. Wind filled sails that slaked a thirst to explore and expand the boundaries of the known world. Arab writings record the use of windmills on the plains of Persia/ Afghanistan from 900 AD. They brought fertility to the land through irrigation and economic wealth from large-scale milling.

Although steam had been used for rotating toys in ancient Rome, its potential for large-scale energy use was not fully realised and exploited until the eighteenth century. The energy source of coal and the prime mover of steam were drivers of the industrial revolution that changed the face of the developed world. Growth in manufacturing and transport built stronger, more expansive societies, and organised labour began to establish safer, more regulated working conditions. The industrial revolution precipitated a period of rapid development of the knowledge and application of energy.

The modern concept of energy was christened *vis viva*, meaning 'living force' in Latin, by Gottfried Leibniz close to the start of the eighteenth century. He defined it as the relationship between the mass and velocity of a moving object. Thomas Young was the first to coin the word 'energy' in 1807, from the Greek word *energeia*, meaning activity, as he grappled with mathematical concepts of mechanics, light and waves. Michael Faraday first induced electricity in a copper coil by passing a magnet through it in 1831 and uncovered one of the world's most flexible sources of energy and energy transportation.

The realisation that all types of energy were interchangeable and conserved came in the 1850s. Less than a decade later, versatile liquid fuel first flowed from oil wells in Pennsylvania. It was the gushing black gold that would soon overtake roughly hewn coal as the primary energy source for the developed world. Within a year, a steamer cut through the ocean from Valentia, County Kerry, to Newfoundland and laid the first transatlantic electric cable to connect Europe to the New World.

The engines of mass transportation were invented from the work of Niklaus Otto, Gottlieb Daimler and Karl Benz at the close of the 1800s. Cars brought transport freedom to the individual and the intense rate of energy output from a plane engine gave us the skies. Populations began to shift with the availability of this energy freedom. A person could fall asleep in one continent and wake in another. The auto engine finally put the last animate energy of faithful farm horses out to pasture. The efficiency of food production for the masses increased, population growth exploded and lands were deserted for the cities. Cities spread from urban cores to encompass vast suburbs from which people commuted on a daily basis through individual and mass transportation means.

What might Henri Becquerel have foreseen when he noticed an innocuous fogging of photographic plates in contact with uranium compounds in 1898? What might Ernest Rutherford have imagined as he pored over abstract drawings of his atomic model and considered its energy potential in 1902? What might have dawned on Albert Einstein when he related the energy equivalence of atomic masses to the enormous speed of light in 1905? Their work led to understanding that the energy of unstable atoms could be released with such immediate intensity that it gave rise to tremendous violence or, when controlled, could be applied to heat and power generation, with beneficial results. None of these physicists changed human propensity for good or evil but their knowledge elevated the consequences of choice to new dimensions of global impact. One such choice led to the horrific end of a world war and the start of a new global tension as powers sought to exploit this brave new atomic energy in order to establish regional or global hegemony.

The information and communications technology revolution started from the objective to process data in a less energy-intensive way in overheated server rooms in California in the 1960s. The speed and efficiency of communication and knowledge transfer that resulted has led to savings in the energy and resource use of governments, companies and individuals and to increasing the emergence of the so-called knowledge-based economies.

Today, the fact that fossil fuels are dwindling and that their negative impact on the environment is significant is widely accepted. This has led to increased efforts to tap energy from natural, renewable sources. The suite of renewable alternatives that can some day fill the carbon-energy void include solar, wind, wave, hydroelectricity, solid biomass, liquid biofuels, hydrogen and geothermal. Even with increased focus in recent decades, renewable energy sources collectively account for less than 2 per cent of the amount of energy used in the world. Yet these sources represent our best hope that the energy history of the future will report how a path of secure and sustainable world growth was chosen during our lifetimes.

TWO

Fossil fuel's fatal flaws

'Rise early, work hard, strike oil.'
– Jean Paul Getty, *My Life and Fortunes.*

It is no coincidence that the greatest period of growth and productivity in human history has occurred during the era of fossil fuels. Unimagined boundaries have been crossed by flight, cars, space travel, communications satellites, agricultural development, organ transplants, computers and genetic sequencing. It has been an unparalleled age for development. In truth, it was the cheap and abundant energy derived from fossil fuels, rather than coal, oil or gas itself, that facilitated these and other developments. If a substitute cheap and abundant energy had been available, it is likely the same developments would have occurred. To date, however, no such substitute has been found or harnessed. Alternatives are now needed more than ever.

Despite fossil fuel's contribution, it is critically undermined by two fatal flaws – it is a dwindling energy source and a devastating pollutant.

THE RISE AND RISE OF OIL

Fossil fuels formed deep in the earth from decayed and packed algae and plants long before the continents split to create the Atlantic Ocean, before dinosaurs lumbered on the surface, before the Himalayas, the Rockies or the Alps were crushed into existence. They lay dormant for millions of years. Here and there through the ages, people stumbled upon fragments of black rock or seeping viscous liquid that they believed had practical and medicinal properties. Ancient Chinese gathered black rocks and oil from surface pools for evening lighting. Babylonians mixed oil and mortar to make cement. Sicilians and Japanese used 'burning water' in clay lamps through the Middle Ages. Native Americans boiled oil and used it for medicine, warpaint and sealing canoes.

Coal played a vital role as the primary energy driver for the industrial revolution, characterised by the large scale application of steam and steel for the first time made possible by coal fire. However, it is the rise of oil, more than any other fossil fuel, which reflects the benefits that can be achieved from access to cheap and abundant energy and also the perils of over-reliance on one energy source.

Oil's use as a commodity fuel came as an innovation borne from necessity. In the 1850s, entrepreneur Samuel Kier operated a handful of shallow brine wells in

Pennsylvania from which he extracted and sold salt. Seeping oil began to pollute his wells and had to be removed. Taking a leaf from local Native Americans, he decided to sell the oil in bottles with labels that read, 'Kier's Rock Oil – Celebrated for Its Wonderful Curative Powers – A Natural Remedy'. The remedy proved popular, despite the noxious contents, but he still had more waste oil than he could sell as elixir. He noted that the liquid was flammable and designed a distillation process, loosely based on that for moonshine, to produce five barrels of lamp fuel a day. His sales grew and attracted others to the area to drill for the new fuel. Colonel Edwin Drake visited Kier's basic operation in 1859 and decided to drill a deeper well through solid granite. At a depth of 69 feet, the drill bit punched through the solid rock and struck oil. He sold his oil for $20 per barrel in the money of the day and started an oil rush.

Within a year, the Pennsylvanian countryside near Titusville was studded with rigs and derricks and the earth was blackened from oil-well blow-outs and gushers. Oil was also struck in Ohio, West Virginia, Kansas and California. So much oil flowed from new wells that whiskey barrels were emptied and diverted to be filled with the black gold. The new industry went from boom to bust within a year. Production reached almost a million barrels of oil in 1860 and the nascent market was initially flooded. Prices tumbled to as low as ten cents a barrel, while cups of fresh water were sold for the same price to thirsty roughnecks at the wellhead.

New markets for oil exploded, from lighting and heating to lubricants. Oil quickly overtook other sources of energy such as coal, wood and whale oil, due to its versatility, abundance and ease of production. The lightest fraction of oil, gasoline, was initially discarded for the lack of a useful application until it was adopted by the newly invented internal combustion engine of the motor car, soon to be mass produced. The natural gas associated with oil was also considered useless and would be vented or flared for a further 100 years. By 1900, 150 million barrels of oil were produced in the US. John D. Rockefeller's highly organised Standard Oil Company held the industry in the grip of monopoly and prices rose. Oil was the new frontier and wildcat prospectors descended on the muddy fields with a fever to find oil, packing Colt six shooters and, as Jean Paul Getty described, 'brawling, bare-knuckled' in the clapboard towns for dominance. Getty, founder of Getty Oil and among the wealthiest people in the world during his life, wrote, 'the American courts ruled that oil is like the wild animal in the jungle. It belongs to the man who finds it and captures it.'

Those individuals who forged the new industry amassed unfathomable wealth. Jean Paul Getty is estimated to have accumulated some €40 billion in today's money, Andrew Mellon, founder of Gulf Oil, €140 billion, and John D. Rockefeller, an enormous $230 billion. Despite many subsequent philanthropic activities, these oilmen were ruthless in their pursuit of oil and money. Rockefeller said, 'I believe that it is my duty to make money.' He also declared that 'the way to make money is to buy when blood is running in the streets'.

Oil exploration had spread to the rest of the world and Russia surpassed the US as the largest producer in the early 1900s. Rockefeller's Standard Oil Company was smashed by the US Supreme Court in 1911 and forced to dismantle into thirty-four separate companies, which included Exxon, Mobil and Chevron. These were joined by Royal Dutch Shell, Gulf, Texaco and Anglo Persian to form what later became known as the 'seven sisters', a cartel-like group that dominated world energy supply until the 1970s.

OIL AS A GEOPOLITICAL FORCE

Oil played a silent hand in shaping the two world wars. In the First World War, Winston Churchill, appointed First Lord of the Admiralty, made a strategic decision to switch the Royal Navy from coal to more agile oil-fired vessels, and consequently maintained Britain's naval supremacy over Germany. In the Second World War, a blockade of oil to the Germans forced them to resort to liquid extraction from coal for transport fuel and ultimately reduced their capacity for troop and materials movement.

Between these wars, the Middle East gained a central position in oil production when the world's largest oil fields were discovered in Saudi Arabia and Kuwait in the 1930s and 1940s and major discoveries were made in Iran, Iraq and the United Arab Emirates. Saudi Arabia became the behemoth among giants and grew to control more than a quarter of the world's proved oil reserves.

At the end of the Second World War, residual forces of the western allies and the Soviets were left sitting on top of the world's most abundant oil reserves in the Middle East theatre of war. The victors undertook a series of negotiations, deals and border recalibrations, in part to facilitate future supply of oil, which moulded the fate of the region for decades. In particular, Franklin D. Roosevelt met with King Abdul Aziz Ibn of Saudi Arabia in February 1945 and agreed to exchange military protection for the steady supply of oil. Billions of dollars' worth of military armaments was shipped to Saudi Arabia in the decades following that agreement and the exchange continues with armaments worth at least $50 billion openly transferred to Saudi Arabia from the US in the 1990s and 2000s.

Middle East countries initially relied on the knowledge and technology of the multinational oil companies to exploit their resources. However, they gradually established their own dominance, in part through the Organisation of the Petroleum Exporting Countries (OPEC), formed in 1960. The oil-producing countries first wrested control from the multinational oil companies by seizing or 'nationalising' their assets. They further flexed their muscles in 1973 when they enforced an oil embargo in response to US support for the Israeli military in its Yom Kippur War with Syria and Egypt. Oil prices quadrupled from $3 to $12 dollars per barrel. Industrialised nations faced immediate stagnation and had no choice but to pay for the energy they required. The oil-exporting countries began to accumulate massive wealth. The cartel baton had passed from Standard Oil to the 'seven sisters' to OPEC. A second crisis in 1979 in the wake of the Iranian revolution sent prices soaring once more towards $40 per barrel.

The dependence of industrialised economies on the politically unstable Middle East region drove early efforts to introduce alternative sources of energy. However, alternatives proved expensive and slow to develop and the industrialised economies reverted to the easier approach of allying with Middle East countries in exchange for continued oil supply and price moderation. The US maintained its bets on Saudi Arabia and Iran, while the Soviets backed Iraq, Syria and Egypt. Vast quantities of arms and funds were shipped to keep the new allies sweet and the oil pipelines open. The US switched allegiance to Iraq during the Iran–Iraq war in the early 1980s, having lost their key Iranian ally, the Shah, who was ousted in the Iranian revolution. Saddam Hussein's regime in Iraq was strengthened with US money and weapons, including chemical and biological agents, as detailed by a US Senate Committee, and this support helped swing the outcome of the Iran–Iraq war. In a further reversal, arms flowed back into Iran from the US, via Israel, in the late 1980s. The US later turned on Iraq and redressed the balance of power in the region over two Gulf wars. The criss-crossing of allegiances and build-up of missiles, military training and money have contributed to the tensions that exist in the Middle East today.

New oil players have emerged from the Caspian Basin, West Africa, South East Asia and Latin America. Nearly every one of them has a history of despotic governance. Industrial economies continue to court or corral them to win favour in the new great game. Never before has there been such an enormous transfer of money and military resources from all of the world's democracies to the world's most anti-democratic and unstable countries, all in exchange for energy.

OIL'S PAY DIRT

The high price paid for oil yields a highly prized return – our way of life. Oil is not only fundamental to economic growth but to economic existence. It is the energy that fuels a vast range of industries, including goods manufacturing, power generation, airlines, chemicals, pharmaceuticals, agriculture, tourism, retail and food – together comprising the entire

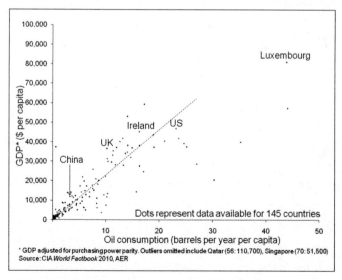

Countries' oil consumption per capita related to GDP per capita

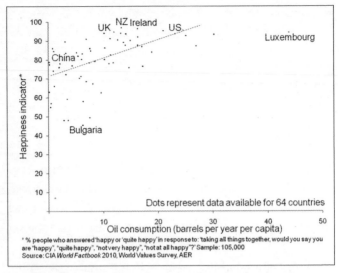

Countries' oil consumption per capita related to happiness

frame of an economy. Moreover, oil contributes to the culture of modern society. It provides the freedom to travel and explore and connects families and communities. It forms the fibres of our luxury goods and the props for our entertainment. It supplies power, heat and light to facilitate neighbourhood gatherings and governing bodies. Without such cheap energy, the current way of life, which new generations tend to assume is rightfully inherited, would not survive.

The relationship between economic performance and the amount of oil consumed by a country can be clearly shown by basic statistical analysis. There is even evidence to show that the average general happiness of citizens in a country is related to the amount oil consumed, most likely due to

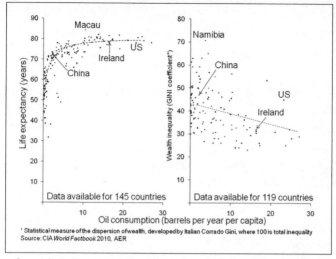

Countries' oil consumption per capita related to life expectancy and wealth inequality

greater access to goods, services and amenities. Similar fact-based analyses show that citizens enjoy better health, longer lives and more equal income distribution (as measured by the Gini coefficient) the more oil their country consumes.

FOSSIL FUEL SUPPLY IS NOW RUNNING OUT FAST

Every finite resource that is consumed must inevitably be exhausted at some point. The imminent end of oil and gas has been signalled by several clear warnings. The average amount of oil and gas discovered each year has been falling for five decades. Over 55 billion barrels of oil were discovered each year in the 1960s. That average had dropped to fewer than 20 billion barrels discovered each year in the decade to

2010. The size of discovered oil fields has also been falling steeply since the 1960s, when the average size of new 'wildcat' fields was 527 million barrels, compared to just 20 million barrels per new field in the 2000s. All of the easy oil and large 'elephant' fields have been found, despite advances in technology for locating new prospects, leaving just small and expensive 'puddles' of oil. Half of the world's twenty largest fields are in decline. The total amount of fossil fuel produced surpassed the amount of new fuel discovered around 1980.

The time at which half of the total resource is used up has arrived. This 'oil peak' is estimated to happen before 2020 and will be followed by a terminal demise when the remaining resource is depleted rapidly by accelerating

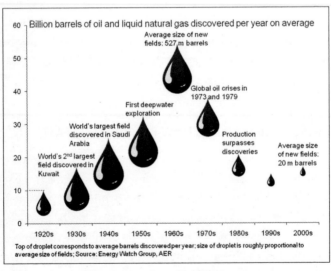

Average yearly global oil discoveries

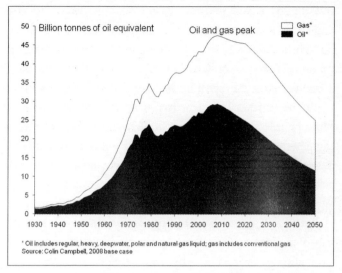

Oil and gas production peaks

consumption. There is some debate about the date of this ultimate peak, with optimists stretching the time to about 2040, while others maintain that the peak has already passed. Every estimate concurs with a common truth: fossil fuels are running out fast. Over 1 trillion barrels of conventional oil have been consumed since it first flowed from Pennsylvanian wells in 1859. The rate of consumption has increased so rapidly in every decade since then that the remaining 1.2 trillion barrels of available conventional oil are likely to be used up in the next twenty to thirty years.

One could think of an analogy where 100 million barrels of oil are sealed in each bottle of stout and a year is compressed into a minute. It is midnight and the party started at 9.30 p.m. So far, ten bottles of stout have been consumed. Revellers drink faster as they warm to the addictive taste and become more animated and five of those ten bottles have been consumed in the last twenty-five minutes. There are only twelve bottles left in the ice-bucket. There is a largely irrelevant discussion about whether there might be another bottle or two deeper in the ice-bucket. More people are piling in the door to the party but in less than half an hour all the black stuff will be gone at the current rate of consumption. The lights are flickering for 'last call', the party will soon be over and the hangover is looming.

Passing the oil peak significantly changes the nature of global oil and gas markets. Falling supply of a resource in the face of rising demand will result in highly volatile prices with alternating spikes and troughs. The overall oil price trend will be upward, despite intermittent price drops. Sudden shortages will be experienced regionally and globally, undermining energy security. Free market competition and substitution will take decades to soothe the supply and demand tension for two reasons. Firstly, primary energy is not a luxury good that has an elastic ability to adapt to supply constraints. Billions of people demand energy for their survival. The primary energy demand of a growing world population cannot simply be scaled back to adjust for lesser supply, even with improvements of energy efficiency and lower consumption behaviours. Secondly, no substitute has yet been fully developed to replace the enormous capacity of fossil fuels to meet our power, heat and transport needs. Most fossil fuel substitutes are still under development, remain relatively expensive, lack the required infrastructure for distribution and suffer the inertia of existing systems to incorporate new energy supply. They cannot alleviate the energy burden for several decades. For instance, it has taken twenty-five years of wind energy development to achieve merely a 1.5 per cent market share of global electricity.

UNCONVENTIONAL OIL WILL NOT HELP MUCH

There are sources of fossil fuels other than conventional oil and gas. It has been estimated that up to 4 trillion barrels of oil equivalent may reside in unconventional sources such as oil shale or tar sands. However, only a fraction of these sources can be practically recovered because of the geological challenges and exorbitant economic and environmental costs for extraction.

Unconventional oil has been produced for several decades near Lake Athabasca in Northern Alberta, Canada, where winter temperatures plummet to −40 °C. Vast areas of surface soil have been scraped away to uncover open-sore pits of oily sand. Additional seams of oily sand are tapped by injecting steam 500 m into the earth to melt the sands and extract the fluid. The sands are upgraded in a process that uses natural gas intensely. It is a particularly inefficient way to produce fuel as the oil from these sands yields only half as much energy as conventional oil for every unit of energy used to produce it. Its production can also emit about 3 times more carbon dioxide than conventional oil. The spent sands are highly alkaline and leaching of the material leads to pollution of streams and groundwater. Dust, poisonous

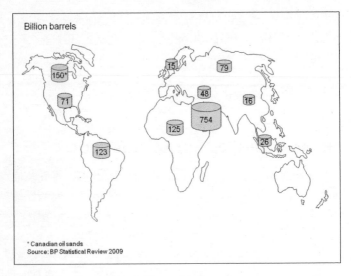

Remaining world oil reserves

hydrogen sulphide, sulphur dioxide, nitrogen dioxide and carbon monoxide are released to the air. Blasting and steam injection render the structure of the subsurface unstable. China, Brazil and Venezuela all possess tar sand resources that are produced under lax environmental requirements. The United States has some 44,000 square km under which oil shale and tar sands exist. The area of highest potential happens to coincide with Rocky Mountain wilderness in Wyoming, Colorado and Utah.

THERE IS NO SUCH THING AS CLEAN COAL

It is estimated that there are some 830 billion tonnes of proven global coal reserves, equivalent to about 4 trillion barrels of oil. This amounts to some 120 years of world supply at current consumption rates. Distribution of the resource is confined mainly to northern temperate zones of the world. It is no accident that the map of coal deposits aligns roughly with the key industrial nations. The largest reserves are found in the US, Russia and China, which together own 62 per cent of proven world coal.

The use of coal as a fuel source peaked around 1910. Oil became the fuel of choice due to its simpler extraction from reservoirs, lower cost and use as a liquid transport fuel. Coal sank deeper into oil's shadow as improved refining increased oil's multiple applications in materials and chemicals production. However, coal is making a comeback as oil runs out and prices increase.

Coal is collected either by strip mining, where a depth of up to 50 m of the earth's surface is ripped away and mechanical shovels eat up to 70 tonnes of coal from a seam with every bite, or by digging shafts and tunnels over 100 m deep to access underground seams. Both methods create significant environmental and safety hazards.

Strip mining completely removes surface plants and wild - life, destroys the soil profile, adds dust to the air and scars the topography even after remediation. Shaft mining claims thousands of lives each year from roof collapse, explosions, gas poisoning or suffocation. For instance, 290 lives are lost on average every year in Ukrainian coal mines. On average 5,080 lives have been lost every year in Chinese mines between 2000 and 2010, as officially reported by the government. The Chinese government also states that 80 per cent of the total 16,000 mineral resource mines in the country are illegal and have no safety regulations. Compensation of £1.5 billion was paid to 65,000 former miners or their widows in 1999 for diseases acquired while working in British coal mines. Diseases included black lung, which results in chronic coughing and pulmonary tuberculosis; emphysema, which results in the destruction of the lung's air sacs, leaving permanent holes in the lung tissue; and chronic bronchitis, which results in long-term coughing, clogged airways, restricted movement and heart disease.

Emissions from the burning of coal include mercury, arsenic, lead, cadmium and trace uranium among other highly toxic elements. The daily radiation from a typical coal-fired power plant is greater than a similar-sized nuclear power plant. Coal burning emits more carbon dioxide than any other fossil fuel per unit of energy. The US coal industry produces more carbon dioxide emissions than all of the road vehicles and planes combined in the nation. The toxic ash by-product from burning coal in the US is dumped in local heaps that cover up to 1,500 acres. The ash contains poisonous heavy metals that can leach into surrounding groundwater and lead to birth defects and cancer in humans and other animals. Most countries do not regulate or monitor the dumping of ash. The US industry alone creates some 130 million tonnes of coal combustion waste each year and dumps it in 1,300 unmonitored sites, many close to large cities, rivers and lakes.

The industry has developed the dangerously euphemistic oxymoron 'clean coal' to describe methods that aim to reduce the disastrous environmental impact of coal mining and burning. For example, gasification of coal enables the controlled extraction of toxic and polluting by-products. The remaining fuel can be used for both power generation and transport. However, the pollutants do not go away. They must be converted or stored deep underground, where hopefully no future geological shift will disturb them. Such 'cleaner' processes are typically 20 per cent more expensive than conventional coal burning. Furthermore, carbon capture devices that collect carbon dioxide emissions from existing smoke stacks result in a reduction of a plant's energy output by 25 per cent.

These approaches do indeed reduce emissions and environmental impact and represent some improvement by the industry. However, they require such large capital investments and high comparative oil prices that they remain at the fringes of industry application, a shopfront of benign aspiration erected by a few developed world companies and governments, while the traditional business of dirty coal continues.

DEMAND CONTINUES TO SOAR

There are two key drivers to increased, almost runaway, use of fossil fuels. The first is the global explosion of population growth. It took from the beginning of man's evolution on earth, some 5 million years ago, until 1950 for the global population to reach 2.5 billion. It took only a further thirty-eight years for an additional 2.5 billion people to join the population count. The growth rate has continued on its exponential trajectory, reaching almost 7 billion by the end

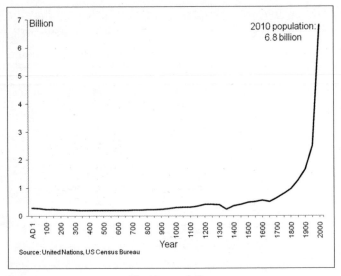

World population

of 2010. Every life has a right to energy for survival. Every new birth increases the demand for fossil fuels – their use in food production and cooking, their application to land irrigation, transport, production of goods and their comfort through heat and light.

The second key driver is the rate of increase of consumption by individuals and societies. Every year, people throughout the world consume more on average than they did the previous year. The long-term trend growth of global energy consumption is some 0.4 per cent per year. This rate is forecast to increase closer to 1.5 per cent per year for the decade leading up to 2020, driven by rapid growth in the developing world as well as changing overall consumption behaviours.

Illustration: Oliver Munday

The 1 billion people in the so-called developed world consume over thirty times more than the remaining almost 6 billion of the developing world. If China and India alone were to catch up on rates of consumption, the global use of energy and resources would triple.

Developing economies claim the right to grow in order to increase standards of living and benefit from the fruits of development that other countries have enjoyed. China shrugged off the 2008–2010 international recession and continued to power ahead with annual GDP growth rates above 8 per cent. Annual growth in sales of consumer goods throughout China remained at a voracious 15 per cent. A fleet of some 50 million vehicles for a population of

1.3 billion is set to swell to over 200 million vehicles by 2020. If China were to emulate the US levels of vehicle ownership of 81 per cent at some point in the future, we would see a further increase of a billion fuel-consuming vehicles on its roads. The Chinese government is rolling out a 30,000 km network of highways, the second longest in the world. China is adding coal plants with a capacity that amounts to eighteen times Ireland's total energy supply every year. The government is also laying claim to energy resources throughout the world with a strong focus on Africa. More than half of public announcements in 2009 made by Chinese premier Wen Jiabao related to energy, including the purchase of international energy companies. In a speech at the World Economic Forum, Jiabao called for 'forceful measures to strengthen the role of domestic demand … especially final consumption … in fuelling economic growth'.

Meanwhile, the developed world continues to use significantly higher amounts of energy and other resources each year. While some new technologies enable activities to be carried out more efficiently and quickly, general industrial and technological advances have resulted in a greater overall use of energy and more people are travelling further and more frequently by road and air. The global advertising industry spends over €200 billion each year shaping the desires of consumers and promoting a belief that energy-intensive goods are not just luxuries but necessities.

Energy-intensive consumption in developed economies is far from efficient and generates enormous waste. For

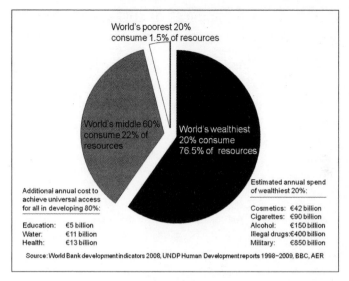

World's poorest 20% consume 1.5% of resources

World's middle 60% consume 22% of resources

World's wealthiest 20% consume 76.5% of resources

Additional annual cost to achieve universal access for all in developing 80%:

Education: €5 billion
Water: €11 billion
Health: €13 billion

Estimated annual spend of wealthiest 20%:

Cosmetics: €42 billion
Cigarettes: €90 billion
Alcohol: €150 billion
Illegal drugs: €400 billion
Military: €850 billion

Source: World Bank development indicators 2008, UNDP Human Development reports 1998–2009, BBC, AER

Share of global consumption

instance, about a third of all food produced in the UK for human consumption using energy-intensive processes is thrown away as waste. This includes about 4 million tonnes of perfectly edible food, worth about €11 billion each year. A survey in England and Wales indicated that about 5,500 whole chickens, 440,000 ready meals and 1.3 million unopened pots of yoghurt are among the good food thrown away every day. More than 150 kg of goods packaging is created and discarded each year per person in Western European countries, about twice the weight of an average person.

ONLY ALTERNATIVE SOURCES OF ENERGY CAN AVOID A SUPPLY CRISIS

Leading world energy agencies forecast that world energy demand will increase by at least a quarter before 2030. The 2010 rate of oil consumption of 86 million barrels per day is likely to rise to about 110 million barrels of oil per day within two decades. However, CEOs of some the world's multinational oil companies, including Shell, ConocoPhillips and Total, have declared that the *optimistic* case for *maximum* oil production is some 100 million barrels a day. The availability of oil will further decline with the reduction in global reserves after the oil peak. It is clear that the post-oil future will require different sources of energy to meet increased global demand. Without alternative sources, at least 20 per cent of global demand would have to be squeezed out of the global appetite, resulting in a highly destructive recession, which would make the 2008–2010 recession seem fleeting and the depression of the 1930s appear rosy. The scramble for dwindling energy resources will create new global fissures and the potential for more war. 'Might' will become 'right' as those nations with the greatest military might will establish their right to energy. Developing and applying alternatives on a vast international scale is the only way to avert the energy supply crunch that is facing the world.

THERE ARE LIMITS TO GROWTH

The growth of world population is currently explosive. Demands for all resources by this expanding population as well as industrial output are also growing at exponential rates, taking long-term trends and even allowing for temporary economic recessions. Modern society appears to be obsessed with growth. Economists and political leaders stress the need to grow. People and countries define themselves and are ranked by their growth characteristics. The desire to achieve growth is like that of a child yearning to mark their height an inch higher on the wall, before they become the adult who is reconciled to the fact that they will not grow taller and focus on cultural, emotional, spiritual or professional growth. It is not true to say that lack of one form of growth means 'the end' – there are many ways to grow.

Unfortunately, material, economic and consumption growth remain the indicators of success for the time being. Their continuing exponential growth in the world sets us on a collision course with unavoidable limits according to a simple principle. Continued exponential growth in demand for any finite resource, such as energy, water, land or air, is mathematically and physically impossible. Growing demand will eventually hit a ceiling.

Examples of some of these limits to growth, described in the late 1960s by an initiative dubbed 'The Club of Rome', include finite fossil fuel energy, the world's estimated 3.5 billion hectares of arable land and the finite amount of available fresh water. If a ceiling is struck without preparation, the forced results are unforgiving. Algae multiplying on the surface of a pool of water will eventually run out of space and use up all of the available nutrients. At that point, the population will collapse, usually to a set of core seedlings.

An intelligent species acting together may be able to adjust to a course of growth that is sustainable and avoids a hard collision with the limiting ceiling.

Part of that adjustment must be to adopt sources of energy other than finite fossil fuel sources. Adopting other finite sources, such as uranium for nuclear energy, provides only temporary reprieve. Only renewable sources can solve the challenge of energy limits. If an energy solution can be achieved, the limiting factor is passed to another resource, such as fresh water, land or clean air. Therefore, new energy sources must be applied in a way that does not propel us towards another ceiling. This presents a further challenge of switching energy sources for continued growth – the definition of that growth cannot include exponential economic expansion and profligate consumption and must complement measurements such as GDP with equally important measurements such as 'collective wellbeing' and 'sustainable use of resources'.

Energy is the first of the key resources to become constrained on a global basis. How we solve for this limit will shape how humanity solves for all future limits we will face.

HOW FOSSIL FUEL ENERGY IS DRIVING CLIMATE CHANGE

GLOBAL ENERGY BALANCE

There is a net global balance of the amount of energy entering and leaving the system of our earth and atmosphere.

Energy entering the system comes from the sun in the form of visible light and ultraviolet radiation, which have short wavelengths. About 40 per cent of solar energy is reflected away or absorbed by clouds and air particles; some 60 per cent of the energy reaches the land and seas and this energy heats the earth. Energy leaving the earth and atmosphere system on the other side of the global energy balance is in the form of heat or long-wave infrared radiation. Some of this emitted heat is returned to earth by a blanket of naturally occurring greenhouse gases, such as water vapour, carbon dioxide, ozone, methane and nitrous oxide. These greenhouse gases help regulate and balance the energy flow and temperature of the earth and atmosphere. However, dependence on these gases to maintain an energy balance leaves the earth's climate vulnerable to changes if the concentrations of these gases happen to change.

At present, the proportion of carbon dioxide in the atmosphere is 0.03 to 0.04 per cent and the proportion of water vapour is about 0 to 2 per cent. They maintain an average global temperature of 35 °C. Without them, the average global temperature would be as low as –20 °C. Fluctuation in the concentration of greenhouse gases in the past can be linked to significant climate changes through history. A direct and striking correlation between global temperature and concentrations of greenhouse gases has been shown for every glacial cycle over the last 400,000 years. The measurement is made by examining trapped air bubbles and isotopes in cores of ancient ice drilled 3 km deep in the Antarctic, which contain a record of past atmospheres.

Past climate change has also been influenced by geological movements and changes to the earth's orbit relative to the sun. Sceptics suggest that human activity can make only a small dent on climate relative to such long-term forces. However, the most recent Intergovernmental Panel on Climate Change (IPCC) report (2007) states that the 'warming of the climate system is unequivocal' and outlines with 'very high confidence' that human activity and greenhouse gas production are direct drivers of climate change. The IPCC is a scientific body of over 400 experts from 120 countries, nominated by governments and multilateral organisations, and its reports are peer reviewed and tested for accuracy by an additional 2,500 international experts. There is no doubt that their conclusions should be taken seriously as the most accurate and robust analysis of climate change and the impact of current energy use. In November 2009, emails at the University of East Anglia in England were stolen by hackers and leaked to undermine the bona fides of a few scientists working on climate change analysis. The emails appeared to contain references that some individuals had played with data to support their conclusions. Furthermore, a mistake about melting Himalayan glaciers was found among the 3,000 pages in the 2007 IPCC report. These were isolated and unrepresentative events. A few sceptics of climate change have seized on the mistakes to question the full scientific body of work. Dr Rajendra Pachauri, chair of the IPCC, suggested that these sceptics were indulging in 'an act of astonishing intellectual legerdemain' and asserted that scientific knowledge was 'something we distort and trivialise at our peril.'

FOSSIL FUEL BURNING IS TIPPING THE GLOBAL ENERGY BALANCE AND CHANGING THE CLIMATE

When carbon compounds are burned in oxygen to release heat or the steam used to drive electric turbines, the residual by-product of carbon dioxide rises into the atmosphere. The burning of carbon in recent decades has been so intense that the addition of carbon dioxide to the atmosphere has exceeded the capacity for natural absorption by vegetation or the sea. The change of land use in certain areas from thick rainforest canopy to urban or agricultural landscape further reduces the natural capacity of the earth to absorb carbon dioxide. In addition, new man-made gases emitted from industrial activity have a greenhouse effect that is far greater than natural gases. For instance, perfluoromethane created from aluminium production and sulphur hexafluoride released from certain cooling fluids have warming propensities that are up to 5,000 and 20,000 times greater than carbon dioxide respectively, although they are produced in small amounts. Other human activity can affect the nature of the global energy balance. For instance, lost energy emitted from cities or industrial areas can create islands of heat radiating into the energy system.

The amount of greenhouse gases now in the atmosphere is greater than at any point in measured history. In particular, the concentration of carbon dioxide is 380 parts per million by volume (ppmv), well above a pre-industrial equilibrium of 280 ppmv in the early 1800s. Some 70 per cent of the increase has occurred since 1970. The last time carbon dioxide levels increased by a similar amount, from 200 ppmv to 280 ppmv, some 10,000 years ago as measured in ice cores, the accompanying increase in the earth's temperature was about 6 °C. As a result, the earth emerged from its latest ice age and sea levels rose by some 120 m. There have been several climatic changes accompanied by natural greenhouse gas variations since that event that have redirected the path of societal progression. A 2 °C drop in temperatures 4,300 years ago led to the collapse of the Old Kingdom in Egypt, the Indus Valley civilisation in India and the ancient Bronze Age in Greece. Similarly, a cold period in the Middle Ages after 1200 AD forced mass emigration and famines across mainland Europe and extinguished Viking settlements that became frozen on Greenland.

Several significant changes in climate have happened in tandem with increases in human-made greenhouse gases. Firstly, global temperatures have risen by about 0.74 °C in the last 100 years. Secondly, rainfall patterns have shifted with increases of 10 to 50 per cent in very high and low latitudes but decreases of 10 to 50 per cent in the tropics and subtropics, which have experienced increasing drought and desertification. Thirdly, global sea levels have increased by up to 14 cm over the last 100 years as a result of melting glaciers and thermal expansion of the seas. Lastly, there has been a recorded increase in the number and frequency of storms, with Europe for instance, experiencing twice as many official storm days during winter and average Atlantic wave heights rising from 2.5 m to 3.5 m in the last fifty years.

Rising temperatures and sea levels are accelerating. Of the last fifteen years, eleven years rank among the twelve warmest in the last century. Temperature has been rising almost twice as fast each year over the last fifty years compared to the last 100 years. Sea levels have also been rising almost twice as fast each year over the last ten years compared to the last fifty years.

CARBON DIOXIDE EMISSIONS GROWTH

Emissions of carbon dioxide have continued to grow strongly despite clear evidence of climate change, international greenhouse gas agreements and even the global economic slowdown. Some 30 billion tonnes were emitted in 2009 from fossil fuel burning, creating a 2.5 cm thick blanket of

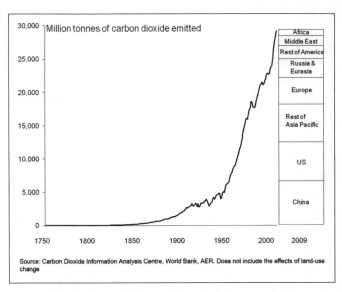

Source: Carbon Dioxide Information Analysis Centre, World Bank, AER. Does not include the effects of land-use change

Global carbon dioxide emissions from human activity

carbon dioxide the size of the entire earth's land area. China has made the leap to first place among global carbon dioxide emitters, surpassing the US in 2006. An additional 4 billion tonnes of carbon dioxide are emitted to the atmosphere each year due to change of land use such as deforestation.

LIKELY CONSEQUENCES OF CLIMATE CHANGE FOR THE WORLD

Climate change measurements and predictions are based on highly complex interactions of ecology, geology, planetary motion and variable human activity over time. The complexity is underlined by the fact that some actions counterbalance others and could contribute to cooling rather than warming. For example, dust raised from industry, car emissions, land clearing and desertification could act as a shield that reflects some of the sun's rays away from the earth, inducing a small cooling effect. Other possibilities are that volcanic ash scattered in the atmosphere from a large eruption could have regional cooling effects or vast amounts of particulates raised from major explosions in a conflict could block sunshine and contribute to a 'nuclear winter'. However, the clear consensus among all but the most hardened sceptics is that unchecked addition of greenhouse gases to the atmosphere at current rates will lead to continual warming and inflict major, damaging consequences on the earth system.

The IPCC outlines the likelihood of a global mean surface temperature rise of 1.1 to 6.4 °C by 2100 and a sea level rise of some 20 to 60 cm, depending on regional location and mitigation activities implemented. Regions in Bangladesh, Egypt, Nigeria and Thailand would experience additional sea level rise of up to 2 m due to delta subsidence, and heavy flooding would affect the lives of over 120 million people. Islands such as the Maldives in the Indian Ocean or the Marshall Islands in the Pacific Ocean would lose more than 50 per cent of land to the waves, rendering them uninhabitable. Monsoon rains and flood patterns would be altered, changing the livelihoods of two fifths of the world's population that live in the monsoon belt. Greater intensity and frequency of storms and the possible increase of El Nino oscillations, the periodic turning of wind and current direction in the Pacific, which have already increased in frequency from 10–15 year cycles to 3–5 year cycles in the last 100 years, would lead to extreme storms, flooding of agricultural regions and forests flattened by wind and fire. Changes to freshwater evaporation and river behaviour would increase the stresses on availability and health of water supplies. The IPCC suggest that freshwater stress would be experienced by 5 billion people by 2025. Despite freshwater shortages, sudden intense downpours and additional warmth would increase the spread of disease borne by mosquitoes to areas as far north as Ireland. Up to 30 per cent of plant and animal species are likely to face extinction if a temperature rise of 1.5 to 2.5 °C is experienced, while up to 70 per cent could face extinction for rises above 3.5 °C.

Global warming may create self-accelerating actions such as the release of large quantities of greenhouse methane from thawing permafrost regions in the high latitudes. The greenhouse effect of methane is twenty-three times more powerful than carbon dioxide. Furthermore, the amount of methane stored in the ground is fifty-five times greater than the amount of carbon dioxide currently in the atmosphere. Warming could also be accelerated due to higher heat absorption by dark patches of land or sea where reflective ice once rested. The tipping point for self-accelerating feedback loops such as these may occur at a temperature that is just 2 °C higher than today. Runaway global warming could result in the major global changes outlined taking effect in sudden jolts over an abrupt ten-year period.

Rising seas, major droughts and floods and food and water shortages would result in untold lost lives and the displacement of large populations. International conflict over scarce resources would inevitably ensue. It is not a comfortable picture. Are the scientists right? The fact-based evidence and analysis is convincing. A certain way to find out is to do nothing.

MITIGATION AND SURVIVAL

Some suggest the best approach to managing climate change is to adapt to changing circumstances as they arise. Raising sea barriers around countries that can afford the infra-structure is an example of this approach in action. It is a brave proposal to negotiate with nature or second guess the consequences of global warming. In its worst manifestation, climate change would represent a metaphorical tsunami that

		Actions that humans decide	
		Take positive action to reduce emissions	Do nothing to reduce emissions
Actions that nature decides in response to rising emissions	Major climate change	Outcome: • Clean planet • Clean-up cost*	Outcome: • Social, environmental and economic meltdown • Extinction of up to 70% of species • Millions of lives lost, populations displaced, famine, flooding, diseases, war
	Little or no climate change	Outcome: • Clean planet • Clean-up cost*	Outcome: • Clean planet

Source: Wonderingmind42 * Worst case scenario is temporary economic recession, best case is break-even

Only one approach in the hands of humans guarantees a clean planet: 'Take positive action'. Doing nothing creates too great a risk.

would overwhelm the social, economic, political and environmental fabric of our existence. The single most responsible way to address the challenge and mitigate climate change is to reduce greenhouse gas emissions and our dependence of fossil fuels as the primary source of energy.

The situation is clear. Our economy, way of life, health, life expectancy and happiness are all dependent on the energy we use. The complication is that we are over-reliant on one source of energy, which is both running out and a serious pollutant. Meanwhile our consumption demand is rising exponentially. This situation has only one possible outcome – global crisis. The only feasible solution is to find alternative sources of energy that are renewable and that are applied with principals of sustainable development. The alternative options that we have are plentiful.

A question of human nature remains: does the strongest human instinct – survival – translate from the individual to the collective group? An individual consistently acts to preserve and protect their own life in the face of a threat. However, when the threat to personal survival is not immediately imminent or apparent to an individual but relies entirely on collective action, will the group act with the same force of instinct? Will lethargy, lack of consensus and confusion shackle the required actions for survival? Such a question has never been tested on a global scale before. The answer will emerge within one lifespan.

THREE

Ireland's voracious thirst for energy

'Nicking and slicing neatly, heaving sods
Over his shoulder, going down and down
For the good turf. Digging.'
— Seamus Heaney, *Digging*.

The need for energy in Ireland has almost doubled over the last twenty years. Ireland's current annual energy requirement is about 190,000 GWh (or 190 billion kWh – see Notes). This demand is divided almost evenly between power, heat and transport. The graph below shows the very rapid increase in demand.

ECONOMIC GROWTH HAS DRIVEN INCREASED ENERGY REQUIREMENT

The primary driver for increased energy requirement was the boom in economic growth that spurred industrial, service

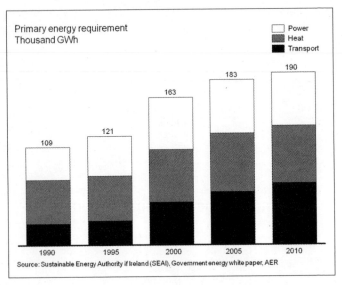

Total energy requirement in Ireland

and consumer activities, particularly over the period 1990 to 2000. Ireland's average annual rate of GDP growth in the 1990s was 9 per cent, three times higher than the European average. Ireland's productivity growth was measured by the OECD to be five times the European average and two and a half times the US average. Dozens of major multinational companies came to Ireland to establish European headquarters, building on the foundations of a highly educated and flexible workforce, attractive corporate tax rates and access to the European Union. US multinationals invested an average of $3.5 billion each year in Ireland and achieved the highest rate of return for investments among all European countries. Irish and world market activities experienced a brief slowdown in 2001 but economic growth soon resumed at more moderate levels until 2008, when it faltered amid the global recession, debt and confidence crises.

A major increase in energy was required throughout this period of economic growth. Manufacturing and industrial facilities applied more heat and electrical energy over longer work days to output more products. More land, air and sea freight was required to transport the products across the country and to export markets. A welcome increase in employment placed hundreds of thousands of additional commuters on the roads. The average spending power of consumers rose, creating additional demand for energy-thirsty production of goods and provision of services.

The sales of new cars rocketed and average engine sizes increased. Adult car ownership grew from 31 per cent in 1990 to 55 per cent by 2007. Annual sales of cars with engine sizes below 1.2 litres fell, while sales of bigger engines above 1.9 litres more than tripled. The demand for transport energy experienced the highest growth rate among all primary forms of energy over the last twenty years. Its share of Ireland's energy mix grew from 22 per cent to 36 per cent.

Increases in population, employment and average personal wealth, accompanied by favourable tax breaks and low interest rates, resulted in a premium market for housing and triggered a construction boom. House prices almost tripled in the ten years before a price peak in early 2007. Some 830,000 new residences were built between 1990 and 2007, roughly half of Ireland's entire housing stock. New

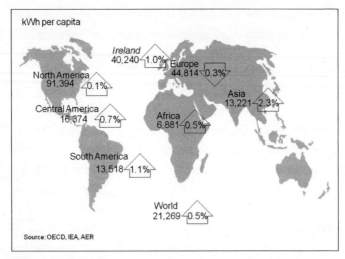

World regions energy demand per capita at the end of the economic boom (2007) and compound annual growth rate (1990–2007)

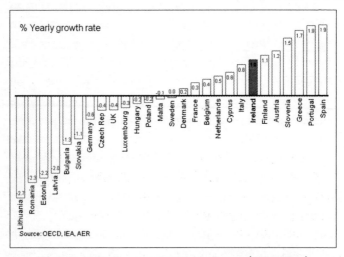

Yearly growth of energy demand in Europe (1990–2007)

house sizes increased by some 24 per cent over that period, while new flat sizes increased by some 27 per cent. The explosion of construction required a major investment of energy and resources, including cement, one of the most energy-intensive products in the world.

The years of rapid economic development resulted in Ireland's demand for energy growing ten times faster than that of the US and experiencing well above average growth in Europe, where energy demand actually fell overall.

ENERGY USE REMAINS AT NEAR BOOM-TIME LEVELS, DESPITE ECONOMIC SLOWDOWN

Economic growth slowed precipitously from 2007. Roughly 33 per cent fewer residences were built in 2008 compared to 2007 and the collapse continued the following year with some 75 per cent fewer new houses registered in 2009 compared to 2008. Average national house prices fell by about 30 per cent in the three years following the housing market peak in February 2007. Over 100,000 construction workers were laid off during that period and national unemployment jumped from 5 per cent to over 13 per cent by 2010, not far from the 16 per cent crisis level experienced in the late 1980s. Car sales fell by over 30 per cent in 2009 compared to 2008 and total household spending was down nearly 10 per cent over the same period. The economic collapse was most pronounced in 2009, when GDP contracted by over 7 per cent, GNP fell more than 11 per cent and 170,000 fewer people were at work compared to the previous year. The economy began to make only tentative progress during 2010.

Despite the market implosion, demand for energy has stayed close to its boom-time level and not fallen in proportion to economic contraction. In fact, despite an economic slowdown of 3 per cent in 2008, energy demand actually rose by 1.5 per cent. The stronger economic contraction in 2009 finally began to dampen energy demand but although the rate of growth has slowed, the overall use remains high. It seems that both the newly built infrastructure and the new patterns of consumer behaviour devour consistently high levels of energy. Overall energy use remains almost twice the level it was before the economic boom and the average energy use per capita is some 40 per cent higher. Furthermore, the pace of past growth was too quick to allow the full adoption of energy efficient behaviours and practices. Many countries in Europe managed to slow or reverse growth in energy consumption partly by using energy more efficiently. Ireland did make some energy efficiency gains but more than 20 per cent of possible efficiencies were lost as a result of increases in energy use due to behaviour. Despite some efficiency improvements in power, more than half of electricity generated in Ireland is lost in transformation and transmission, a rate that is higher than international averages. A legacy from Ireland's boom times is a voracious thirst for energy.

Demand for energy in Ireland continues to rise. Energy requirements could increase by as much as 25 per cent by 2020 if there are no improvements in energy efficiency or new renewable energy projects are not introduced. The government has set a target to achieve a 20 per cent increase in energy efficiency by 2020 to mitigate this potential increase in energy requirements. It has also set targets for increased contributions from renewable energy (40 per cent for power, 12 per cent for heat and 10 per cent for transport). However, even if these targets are met, overall energy use could still increase by several per cent. In either scenario, Ireland continues to catch up with levels of consumption in North America as it slakes its new-found thirst for energy.

WHERE DOES IRELAND'S ENERGY COME FROM?

Fossil fuels account for some 95 per cent of energy use in Ireland. Furthermore, nine times more energy is imported than is produced at home. This creates a dependency on foreign fossil fuel that threatens economic stability, national security and environmental sustainability. The amount of yearly imported energy more than doubled in the period 1990 to 2008, while energy produced in Ireland has fallen by more than half over the same period.

Imported energy is comprised primarily of oil, gas and coal. Imported oil accounts for over 96 per cent of Ireland's liquid transport fuel in the form of petrol, diesel and kerosene. The remaining 3–4 per cent is mainly biofuel, more than half of which is also imported. Imported oil, gas and coal account for 80 per cent of electricity generation fuel and 88 per cent of primary heating fuel, with peat,

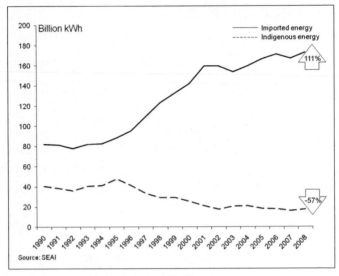

Irish imported and indigenous energy

indigenous gas and renewable sources such as wind making up the balances. Most countries have some reliance on imported fuel but the level of imports and the singular reliance on hydrocarbons makes Ireland's dependence a major comparative weakness.

A serious oil shortage shock, an international fuel dispute or an incident relating to the two parallel gas pipelines that stretch from a single location in Scotland to Ireland could have an immediate and catastrophic impact on Ireland's industry, commerce and the livelihood of every individual. Immediate rationing would be enforced on the 90-day national oil reserves stored in tanks on Whiddy Island in Bantry Bay and at selected port terminals. Industrial activity

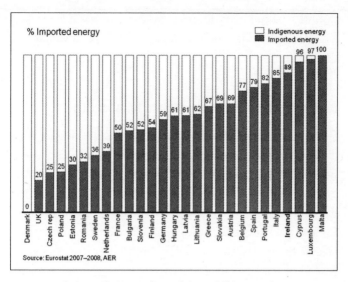

% Imported energy

Indigenous energy
Imported energy

Denmark 0, 20
UK 25
Czech rep 25
Poland 30
Estonia 32
Romania 36
Sweden 39
Netherlands 50
France 52
Bulgaria 52
Slovenia 54
Finland 59
Germany 61
Hungary 61
Latvia 62
Lithuania 67
Greece 69
Slovakia 69
Austria 77
Belgium 79
Spain 82
Portugal 85
Italy 89
Ireland 96
Cyprus 97
Luxembourg 100
Malta

Source: Eurostat 2007–2008, AER

Imported energy dependency of EU countries

would be severely constrained and national power shut-downs could be imposed. Businesses would close and services would shudder to a halt. Long lines would form at petrol forecourts for limited transport fuel, similar to those experienced during the oil crisis of the 1970s. Commuting to work would become difficult for all except those within walking or cycling distance. Heating fuel prices would soar, deliveries would be cancelled and people would resort to gathering sticks for the fireplace. Any prolonged shortage of energy resources would create a deeply destructive economic depression. An atmosphere of helplessness would prevail. A sudden call for national resilience and self-sufficiency would fall flat on the 'too late' realisation that the natural energy resources at our own doorstep were out of reach for lack of facilities and infrastructure required to take advantage of them.

DWINDLING INDIGENOUS FOSSIL RESOURCES

Fortunes have been invested and lost on searching for Irish oil over many decades. Over €2 billion has been spent drilling 160 oil exploration and appraisal wells in Irish waters but no viable oil fields have ever been found. Small gas discoveries have been made at the Kinsale and Corrib fields. Kinsale gas in the Kinsale Head, Ballycotton and Seven Heads fields lies some 25 km from shore, under about 100 m of sea water and a further 1,000 m of rock beneath the seabed. Kinsale gas was discovered in 1973 and first piped ashore in 1976. Production peaked in 1995 and subsequently declined rapidly, falling by 84 per cent in the following fifteen years. The Corrib gas field was discovered in 1996 in the Slyne basin, 80 km west of Mullet Peninsula, County Mayo, under 355 m of water and a further 3,500 m beneath the seabed. The gas is estimated to have a lifetime of just fifteen to twenty years.

Peat has long been a source of local fuel in Ireland. Rough slabs of moist, 3,000-year-old peat bog have been cut by hand and stacked to dry before burning since the 1700s. Up to half of Ireland's peat resource was used up in this way by 1946. In that year, the Irish government established Bord na Móna to accelerate the depletion of peat by mechanical cutting and to apply the peat to heating, electricity generation and horticulture. However, peat production from the available resource has declined by 65 per cent since 1995. Furthermore, Ireland's commitments to the Kyoto protocol will result in a continued reduction in use of indigenous peat as a fuel, due to its high carbon dioxide emissions when burned.

THE JOURNEY OF IMPORTED OIL AND GAS

Oil and gas are the primary sources of imported fuel for Ireland's power stations, home burners and vehicles. They account for a staggering 61 per cent and 28 per cent of imported energy respectively. These life-giving yet lethal fuels have been drawn mainly from the UK and Norway, subject to previous fuel trades from as far afield as the Niger Delta, the Middle East, Russia and the Gulf of Mexico.

The fuel is traded on international commodity markets, which thrive on volatility. The markets are like nervous animals that flinch at any stimulus. The price of fuel is sensitive to international relations, daily business conditions, casual remarks by economists, harsh rhetoric from politicians, behavioural whims of consumers, high seas and weather forecasts, among other interacting factors. The greatest measure of success and safety in trading and transporting oil or gas is that it is never once seen on its entire journey from ancient reservoir to final combustion chamber.

Deposits of oil and gas begin life in pockets of subsurface rock, once the layered beds of decaying plants and animals.

They have lain trapped in deep reservoirs by impermeable rock caps for an average of 35 million years. Their final journey begins when engineers in luminous overalls sitting in trucks or Portakabins discharge tiny seismic tremors through the earth and record the reflections in a three-dimensional picture of the subsurface. When a likely reservoir is identified, a drill bit snakes its way from a surface rig and can be guided along a winding trajectory around hard rock formations several km deep and as far as 10 km horizontally displaced from the point of entry to reach a target within 1 m of accuracy. When it taps into a reservoir, the fluid is siphoned and tested and if the results are successful, a temporary plug is inserted. Then steel and concrete well casing is set along the drilled hole walls to stabilise the borehole. The plug is removed and the hydrocarbon is drawn to the surface processing equipment. It may gush under its own pressure or be coaxed by the sucking action of a 'nodding donkey' or the injection of steam. Toxic heavy metals and the most deadly gases are removed at the surface. Finally, the fuel is piped or shipped towards its market.

Supertankers as long as the Empire State Building is tall carry up to 300 million litres of crude oil to Rotterdam for the European market. Here the ships break bulk for onward distribution in smaller parcels. Delivery can also be made directly to oil refineries that feed the Irish market at Pembroke in Wales, Stanlow near Manchester, Mongstad in Norway or Whitegate in Cork Harbour. At the refinery, the bulk crude oil is boiled at some 600 °C until it turns into gases. Crude oil's different components cool from their gaseous state back to liquid at different temperatures and therefore can be separated when passed through a column with a varying temperature from bottom to top. Heavy oil fractions turn back to liquid first with falling temperature. Lubricating oil and heavy gas oil cool to liquid and are separated at high temperatures, while petrol and kerosene are separated at lower temperatures. The heavy oils are further cracked or altered into useful chemicals, plastics and fuels.

The refined oil fractions are pumped into smaller ships and trucks for the next stage of their journey. Some 65 per cent of Ireland's refined oil fuel arrives at Dublin Port and the balance passes through Cork Harbour. The fuel wallows in squat dockland storage tanks that hold up to several million litres until a road tanker that can carry some 25,000 litres drives up to the dispensing gantry and fills up. At last, the fossil fuel is unloaded into an underground storage vessel at a roadside forecourt. Here it waits to be pumped its final few feet into a vehicle tank through a 1-inch nozzle, gripped firmly by a consumer, at a rate of up to 80 litres per minute. The consumer is likely to be thinking about anything except the journey made by the fuel that is rushing through the nozzle in her or his hand.

A third of European gas gushes across 5,500 km in pipelines beneath vast fields of Russian corn, the former Soviet planned cities of Belarus and Ukraine, industrial valleys of Germany and dinner tables where Dutch families unassumingly gather to break bread.

The gas travels at about 35 km per hour and its fine particles of grit and corrosive chemicals eat away at the carbon steel pipes like sandpaper and acid, so much so that mechanical pigs are launched down the pipes to test and recoat the insides at regular intervals.

Gas bound for Ireland hurtles across the North Sea and through Scotland. Just two parallel pipelines, of 24-inch and 30-inch diameters, stretch to Ireland from Moffat in southwest Scotland. They enter the sea at Ross Bay, Scotland, skirt the Isle of Man and make land at Dublin and Meath respectively. A mix of Norwegian, British and Russian gas is directed to Waterford, Cork, Limerick and Galway as needed.

It can take just several weeks for the multimillion-year-old oil or several days for gas to be wrenched from its reservoir and dispensed at the petrol pump, dumped into power station bunkers or flushed into a back-garden home-heating tank. Its final combustion occurs in a flicker. A spark or flame is applied to the fuel. It combusts with oxygen and its atoms instantaneously collapse to a lower state of energy. Heat is released in the process and can be put to work to drive our electricity turbines, our car engines or our home heating – all contributing to the constant background hum of the daily activities we take for granted.

IRELAND'S GREENHOUSE GAS EMISSIONS

Irish greenhouse gas emissions approaching 2010, equivalent to some 70 million tonnes of carbon dioxide, were almost 25 per cent higher than 1990 levels. Although there has been a minor fall-off in emissions due to a partial upgrade of

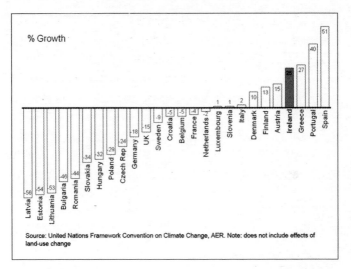

Source: United Nations Framework Convention on Climate Change, AER. Note: does not include effects of land-use change

Total growth of greenhouse gas emissions in Europe (1990–2008)

electricity generation to gas-fired stations and the cessation of fertiliser production, Ireland's greenhouse gas emissions remain over 10 per cent above its agreed Kyoto target and are projected to rise a further 10 per cent by 2020 if no additional reduction measures are implemented.

WHO ARE THE BIGGEST EMITTERS IN IRELAND?

There are 105 energy-intensive carbon dioxide emitters officially registered in Ireland to avail of carbon emission allowances. These include the largest commercial sites for electricity generation, cement production, alumina refining, oil refining, peat briquetting, dairy processing, alcohol brewing and other general industry activities. Together they

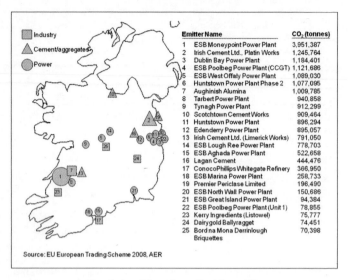

	Emitter Name	CO$_2$ (tonnes)
1	ESB Moneypoint Power Plant	3,951,387
2	Irish Cement Ltd., Platin Works	1,245,764
3	Dublin Bay Power Plant	1,184,401
4	ESB Poolbeg Power Plant (CCGT)	1,121,686
5	ESB West Offaly Power Plant	1,089,030
6	Huntstown Power Plant Phase 2	1,077,095
7	Aughinish Alumina	1,009,785
8	Tarbert Power Plant	940,858
9	Tynagh Power Plant	912,299
10	Scotchtown Cement Works	909,464
11	Huntstown Power Plant	896,294
12	Edenderry Power Plant	895,057
13	Irish Cement Ltd. (Limerick Works)	791,050
14	ESB Lough Ree Power Plant	778,703
15	ESB Aghada Power Plant	522,658
16	Lagan Cement	444,476
17	ConocoPhillips Whitegate Refinery	366,950
18	ESB Marina Power Plant	258,733
19	Premier Periclase Limited	196,490
20	ESB North Wall Power Plant	150,686
21	ESB Great Island Power Plant	94,384
22	ESB Poolbeg Power Plant (Unit 1)	78,855
23	Kerry Ingredients (Listowel)	75,777
24	Dairygold Ballyragget	74,451
25	Bord na Mona Derrinlough Briquettes	70,398

Source: EU European Trading Scheme 2008, AER

Ireland's top 25 carbon dioxide emitters

produce about 20 million tonnes of carbon dioxide each year. The top 25 sites in this group produce 94 per cent of emissions in commercial industry.

Every individual is responsible for some greenhouse gas emissions. At a small level, each person exhales more carbon dioxide than they inhale as they combust fuel in their bodies. A jogger exhales about 2 g of carbon dioxide for every km she or he runs, while a cyclist exhales about 0.5 g over the same distance. The combustion of fossil fuels in machines and vehicles releases greenhouse gases in a more intense way. An average family car releases in the region of 200 g of carbon dioxide for every km driven. To give a sense of physical scale, a 5 km journey in a family car releases enough carbon dioxide to fill the capacity of a bath, where the gas is at standard temperature and pressure. Choosing to travel by city bus, commuter train or DART instead of a car in Ireland reduces the carbon dioxide emissions per passenger by a factor of four, assuming each mode of transport is full with passengers. Air travel releases as much as some 50,000 g (50 kg) of carbon dioxide per km. The amount of carbon dioxide released on a return flight from Dublin to New York per passenger is more than that released by an average family car over a whole year.

Agriculture and animal farming are also major sources of greenhouse gas emissions. Cows in particular emit methane gas at a phenomenal rate. A cow can emit 120 kg of methane in a year – over a thousand times more than a human. The greenhouse gas effect of a cow's annual emissions is equivalent to driving a family car for 15,000 km. Sheep emit about half the methane of cows, and pigs and chickens emit about one-seventh less, per unit weight of each animal.

LIKELY CONSEQUENCES OF CLIMATE CHANGE FOR IRELAND

The effects and consequences of forecasted climate change in Ireland are dramatic. An initial surge of amusement that current temperatures would increase and summertime rainfall decrease is replaced by a fearful jolt from the realisation of the more likely negative impact on life.

A rise in global sea levels of some 60 cm would result in large tracts of land being subsumed by coastal flooding and increased wave attack. Every 1 cm of sea level rise will force land to retreat by 1 m at low-lying coastlines. A total of 60 m of land could be claimed by the sea at all sandy coasts. Some of the sea level increase may be counterbalanced by the fact that Ireland's land mass is still rebounding slightly since the disappearance of its ice cap. However, several hundred square km of Ireland's total land are at risk of being lost to sea encroachment as well as wave and tidal surges and more frequent storms. Farms, railway bridges, airports and cities and towns such as Dublin, Cork, Galway, Limerick, Shannon, Dungarvan and Wexford are under threat and face a damaging surge in water levels.

Temperatures in Ireland are rising three times faster than the global average. Average winter temperatures are projected to rise by some 1.4 °C by 2050, accompanied by about 10 per cent more rainfall. Floods are likely to increase in intensity and regularity and cover more land. Towns in flood plains and river valleys are at significant risk of repeated flooding and damage to businesses and homes. The defence forces assisting evacuation for hundreds of residents in Fermoy, Mallow, Bandon, Skibbereen, Clonakilty, Ennis, Clonmel, Athenry, Ballinasloe and scores of other inland towns will become a regular feature. Flooding will increase the risk of drinking water contamination from animal manure, human sewage and fertiliser for bigger towns and cities, particularly in the west.

A projected increase in average summer temperature by about 1.5 °C by 2050 and a decrease in summer rainfall by some 15 per cent are likely to lead to droughts, interspersed with intense downpours, particularly in the east and south-east. The surface of peatlands may become temporary dustbowls in the summer, releasing carbon to the air and acid to groundwater. Such drought will render potato growing less viable across large parts of the country although overall yields of barley, maize and forest trees may increase. Stronger blooms of marine algae and warmer water are likely to suffocate salmon and fish farming in the west and southwest. The birth and survival rates of rats, bats and cockroaches is likely to increase and new species of migrant insects as well as birds are likely to settle in Ireland.

The potential effects of climate change in Ireland are profound. Actions both to reduce the potential for change and to prepare for possible changes that may occur despite our best efforts are needed. However, Ireland's actions alone cannot mitigate a threat that is determined globally and felt locally. The most powerful action that Ireland can undertake is to lead by example. Ireland has a unique opportunity to achieve energy independence and reduce emissions by choosing energy alternatives and managing energy efficiently. This can result not only in a more secure, healthy and prosperous society but can have a lasting global impact through the act of leadership.

FOUR

Drinking water from a fire hose: alternative energy options

What are the best alternatives to fossil fuels for the three key energy uses of transport, power and heat? There is no silver bullet solution – no one alternative that will fill the void left by finite fossil fuels. The luxury one-stop shop of hydrocarbons for our energy needs has been almost consumed in the blink of a couple of centuries.

Most of the energy that exists in our world has arrived and continues to arrive from the sun. Sunshine provides light for plants to grow, warmth for the planet and is responsible for weather, climate and the cycle of water in rain and rivers. Therefore, the sun is the source of most of the key options for energy in the world, namely wind, wave, biomass, hydroelectricity and of course direct solar heating and power.

The remaining options include significant energy that can be released from atomic nuclei, gravitational energy exerted on the earth's seas by the moon, sun and other planets, creating tides, and seeping warmth from the earth's core that provides pockets of geothermal energy.

Each alternative to fossil fuels has distinct benefits, challenges and trade-offs. The focus here is on those alternatives that are most abundant and have the strongest potential in Ireland. The 'big five' sustainable options for Ireland are wind, solar, wave, biomass and managing energy efficiently. The nuclear option is a further clear alternative that will be explored, though is not truly sustainable as nuclear fission relies on a finite feedstock.

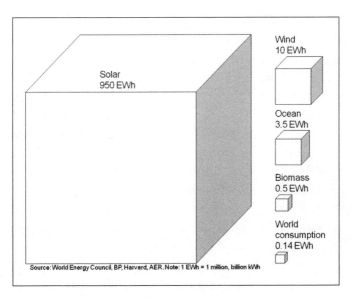

Global annual energy resources

The schematic above illustrates the abundance of alternative energy resources on a global scale. The sub-chapters describe how these resources can be applied both at a world level and in Ireland. The sub-chapters are sequenced in order of the total long-term potential impact for Ireland of each resource.

WIND

'On a day when the wind is perfect
The sail just needs to open
And the world is full of beauty.
Today is such a day.'

– Mowlana Jalaluddin Rumi,
On a Day When the Wind is Perfect.

SHAPING THE WIND

The air we breathe is made up of 78 per cent nitrogen, 21 per cent oxygen and small amounts of argon, carbon dioxide, ozone, noble gases and water vapour. The blanket of air closest to the earth's surface, called the troposphere, extends about 12 km upwards and is where all the key actions of weather take place. This air moves in two planes, vertical and horizontal. The horizontal movement of air is called wind. Wind can race up to 1,000 times faster than the slow vertical sinking and rising of air.

Wind movement is governed by four key forces – pressure differences, gravity, the earth's rotation and friction. Pressure differences result from the uneven distribution of heat energy from the sun. The greater the heat energy, the more the gaseous particles in air bounce off each other and the greater the pressure. Air naturally moves from areas of high pressure to areas of low pressure. The greater the difference in pressure, the faster the wind. Areas of low pressure, or depressions, on a weather chart often herald cold air, while areas of high pressure, or anticyclones, can usher warmth.

Warm air is less dense than cold air, so it rises, just as happens with a hot air balloon. As it rises, the air particles spread out and the pressure and temperature decrease. The void left beneath the rising air will be filled by air breezing in from a different location. This happens for instance when warm air near a coast rises from the land and a cool breeze is drawn from the sea. Rising air will be met by a counterbalancing force as gravitational energy is exerted by the earth on air molecules, pulling them slowly downwards again.

As the earth rotates, it effectively moves beneath the suspended air. An air particle that rose up under pressure and later sank again would find that the earth had moved on in the interim and that it was displaced from its original starting point. This effect, the Coriolis deflection, contributes to the patterns of wind.

Finally, the changing profile of the earth's surface creates a frictional drag on air moving over it. Friction can affect both the speed and direction of wind. Wind can be gustier as it passes over hills or is thrown off course by whirling eddies of rising warm air. Winds at sea and at high elevations can move faster due to the absence of obstacles causing friction.

Wind speed is consistently measured at a height of 10 m by meteorologists to allow for low-lying frictional effects and also to ensure measurements can be compared like for like across the world.

The effects of these four forces fluctuate across geography and over time and combine to create the turbulent complexity of wind. They result in distribution of warm and cold air around the planet and transportation of life-giving water vapour. The kinetic energy of wind can perform work and be converted into other forms of energy.

ROMANTIC WINDS

Wind is so innately familiar, accompanying us in every outdoor moment, that we personify it in name, character and romance. The sirocco blows north over the Sahara, carrying sandstorms that can blot out the sun; the chinook whispers down the easterly flanks of the Rocky Mountains into the plains of Montana, melting the snows. The mistral is a blustery, violent wind that gusts south along the Rhône Valley to the Côte d'Azur and Mediterranean in winter and early spring. The mistral is known as a valley wind. It forms as chilled air sinks slowly down the sides of Alpine valleys. It settles in wallowing pools of cold air, waiting for the trigger of a pressure change to send it spilling out of the valley and down the twisting ravine in bitter gusts to the sea.

The Trade Winds are formed about 30° to 35° both north and south of the equator where heat energy is relatively intense. These winds start to rush towards the equator and are deflected in an easterly direction by the rotation of the earth. When they meet each other close to the equator, they push upwards leaving a void below called the doldrums, where light winds and low pressure can result in a 'dead calm'.

The Roaring Forties are boisterous winds that blow west to east beneath the 40° latitude line in the southern hemisphere. This was the dependable wind channel into which clippers once ventured from the Cape of Good Hope to race across the Indian Ocean to Australia.

Revolving tropical storms are known as hurricanes in the North Atlantic, typhoons in the Pacific, cyclones in the Indian Ocean and willy-willies north of Australia. Localised whirlwind storms over land are known as tornadoes or twisters. Wind speeds can rage at 400 km per hour and wreak devastation in their paths.

Occasional 'Showers of Blood' occur as sandstorms in the Sahara which whip red dust high into the air. The dust may be lifted as high as 5 km and then carried by winds as far as Ireland, where rain washes the desert sands earthward and covers the land in a red Saharan veil.

CAPTURING WIND ENERGY: HOW DOES IT WORK?

Although wind patterns are complex, the average wind speed in a specific location over a full year is predictable. Therefore, the average energy available from wind can be mapped and sites with the highest potential can be considered for the placement of wind turbines that convert the kinetic energy of wind into electricity.

The largest commercial turbines have blades that are over 60 m long and sweep an area in the air the size of a football field. The total height of these turbines can be as high as 200 m with the blade extended, which is roughly the height of a sixty-storey building. Smaller commercial turbines are typically some 70 m tall, with a 25 m blade extended. Turbine towers are usually made of steel and the blades made of a fibre laminated plastic. Micro wind turbines can be just a few metres tall and can be attached to the top of a domestic roof to provide local power.

The kinetic energy of wind intercepted by the aerodynamic blades of a wind turbine makes them rotate. The blades are connected to a shaft which provides motion to magnets in a generator, usually with the help of gears. These turning magnets induce the movement of electrons in conducting cables and the generation of clean electrical energy begins. A transformer increases the voltage of the electricity to that required for efficient transmission.

Large wind turbines are set to 'cut in' at feasible wind speeds of about 12 km per hour. They also 'cut out' at high speeds that could make the structure unstable. Large turbines can manage wind speeds of up to nearly 100 km per hour and blades can spin as fast as 25 revolutions per minute. Wind energy increases rapidly with wind speed as it is proportional to the cubed velocity of the wind. For instance, a 10 per cent increase in wind velocity results in a 33 per cent increase in available energy.

Advances in design have enabled turbines to increase maximum power output sixfold over the twenty-five years to

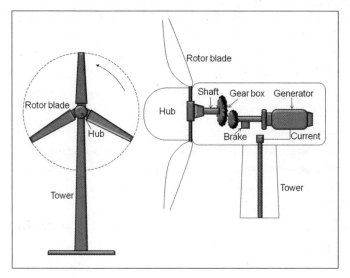

Wind turbine schematic

2010 and performance continues to improve. A single large turbine with an energy capacity of 6 MW (million W) can produce enough electricity for 3,000 European households. Micro wind turbines can produce in the range of 1 to 6 kW (thousand W), enough to power a home or small business. This electricity generation has the benefit of producing no greenhouse gases while operating, although some fossil energy will be used in the initial turbine manufacture. Turbines also have the advantage of having a relatively small footprint in terms of actual ground area.

The noise of air rushing past the blades at the base of a wind turbine can be some 50 decibels. This is about the same as a typical office environment or the inside of an

urban house. The noise reduces with distance and is camou-
flaged by the pervasive sound of the wind itself. When there
is no wind, the blades fall silent. There is some evidence
that, despite these relatively low levels of noise, the sound of
wind turbines along with light flicker from turning blades
may affect the health of people living very close to wind
farms. Some residents living 'in the shadow' of wind farms
have reported headaches, higher levels of anxiety and forms
of depression. This underlines the importance of careful
environmental and social impact assessment and planning
for any wind farm development.

Wind is a capricious source of energy. Although the
average wind energy at a location can be accurately measured
and forecast over the course of a year, wind blows inter-
mittently and at different speeds on a day-to-day basis.
Energy output from a turbine may be possible less than a
third of the time as a result. This wind energy productivity,
or load factor, compares with load factors of some 65 to 85
per cent for coal, gas and nuclear power plants. Wind energy
supply must therefore be accompanied by other sources of
supply in order to provide a smooth and consistent output.
A means to overcome the intermittent nature of wind energy
and provide consistent supply is to store the energy as it is
captured. For instance, wind energy can be used to pump
water to a higher elevation, where it can later be used to
generate hydroelectricity on demand. The energy can also
effectively be stored by using the immediate electricity generated to produce hydrogen or compress air, both of which can
be transported and put to work at a later time as needed.

HOW MUCH WIND ENERGY IS AVAILABLE?

Just over 1 per cent of the sun's radiation that reaches the earth is converted to wind energy through the effects of heating and cooling. This equates to an available energy of over 10,000 PWh (or 10 million, billion kWh) per year, seventy-five times the world's total energy consumption and 590 times the world's electricity consumption at present.

However, several natural constraints reduce the amount of energy that can actually be harnessed. Wind energy at locations far out to sea and those covered by ice, forests and urban landscape as well as areas with low average wind speeds can be excluded. Furthermore, wind turbines efficiently convert up to only 60 per cent of wind energy when it is blowing, have an overall efficiency of around one third and turbines also require spacing to minimise airflow interference from other turbines. The energy that is available to be captured is about one fifteenth of total wind energy when these aspects are taken into account. It is 690 PWh (or 690,000 billion kWh) per year, which is five times current global total energy consumption or forty times global electricity consumption.

The world's current electricity consumption could hypo - thetically be met by 200,000 offshore 6 MW turbines and 1.6 million onshore 3 MW turbines, all working at a third of their capacity. They would take up less than 0.1 per cent of the earth's surface area, including 0.3 per cent of land area, allowing spacing of about 0.25 square km per turbine. They would cost in the region of €8 trillion. This cost is equivalent to 0.5 per cent of global annual gross domestic product for twenty-five years.

These facts provide a sense of the order of magnitude required to tap the power of wind for clean energy that is almost free once turbines are operating, allowing for maintenance costs. However, the actual introduction of wind turbines is limited by the inertia of human and resource systems. Project planning, raising finance, lead time for both manufacturing and installation as well as creating infrastructure and access for electricity distribution all take time. Furthermore, wind energy will only effectively substitute other energy sources if it can compete economically. Wind-generated electricity can cost in the range of 75 to 200 per cent of the cost of fossil fuel derived electricity, depending on the location.

The current installed global wind energy capacity is just 1.5 per cent of total world electricity capacity. Despite the nascent state of the worldwide industry, 400,000 people are employed and €40 billion in annual revenue is generated in the sector. The continuing fall of wind energy costs and the long-term upward trend of oil prices mean that wind energy will improve its cost competitiveness and play an increasingly important role in energy supply. Global wind energy capacity is growing at an explosive rate of 25 per cent each year with the fastest growth experienced in the US, China and India. It is estimated that as much as 35 per cent of global electricity will be met by clean wind energy by the middle of the century.

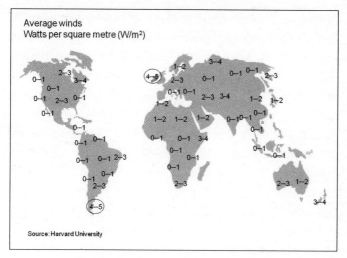

Global average wind energy rates available

WIND ENERGY IN IRELAND

The Big Wind

Sunday 6 January 1839 started quietly. The wind was still and there was unseasonal warmth in the air. This was the Christian and Gregorian feast of revelation, the twelfth day of Christmas, and the eve of the pagan day of judgment.

In the Atlantic to the northwest of Ireland, one of the deepest depressions ever known to Irish or British meteorology was forming and creeping slowly towards land. Mercury plummeted below the bottom gauge of barometers throughout the country. The advancing depression collided with a blanket of European warm air. The warm air was pushed up through the atmosphere and cold winds from the North Atlantic rushed into the void at a ferocious speed. By the time the weather front hit landfall at the west of Ireland, the winds had been whipped to super-storm force and unleashed a furious battering on anything in the tempest's path.

Contemporary news journals would later describe the devastation. A thunderous rumbling in the west preceded a rushing blast of winds that swept across the country, carrying destruction and plunging it into darkness. Cod and eels were blown out of the seething waves and left flapping in the dunes and fields of Mayo; gulls were blasted against the cliff faces in Galway and fell in heaps on the rocks below; stones were plucked from earthen banks and struck sheep dead in Clare. Tens of thousands of rural homes collapsed; falling walls and chimneys crushed and killed an estimated 200 inhabitants across the counties; grown men were lifted up and smashed into the ground; small animals were swept into the air and disappeared forever; fires raged as the embers of exposed hearths hurtled through the air; tethered animals were strangled or roasted alive. The Shannon and Grand Canal burst their banks and flash floods carried whole families to their death; vast swathes of forest trees were cut down like reeds; and tombs in graveyards burst open. The cities of Dublin, Belfast, Limerick and Cork were sacked by the winds. People were forced to crawl on their hands and knees through the streets and hide in drains as rooftops were prised from their homes, glass shards and slates flashed through the air and fires spread through houses and churches. Over twenty ships were dashed by the winds on to rocks and lost, including the Irish emigrant ship, the *Lockwoods*, bound for New York, with the loss of fifty-two lives, and the transatlantic liner, the *Pennsylvania*, bound for Liverpool.

This was the Night of the Big Wind (or 'Óiche na Gaoíthe Móire' in Irish). There had been other great windstorms in Ireland and many more to come but this one found a place in folklore and epitomised the devastating energy of the wind when it is concentrated in time and place. The harvest of this vast source of energy for the benign purpose of fuelling a nation has only just begun.

Wind energy potential in Ireland

Ireland is blessed with some of the best wind resources in the world. The average wind strength and consistency on this windswept island facing the Atlantic southwesterlies is an enormous natural resource that is waiting to be tapped. The total wind energy for Ireland is estimated to be at least 6 million GWh (or 6 trillion kWh) per year, based on wind speeds and area available. To give a sense of scale, this is nearly 200 times the current total electricity consumed in Ireland.

However, some 380,000 GWh (or 380 billion kWh) of wind energy per year is more feasibly accessible if areas with low average wind speeds and inaccessible locations are excluded and when the productivity of current wind technology, which is about a third, is taken into account.

This is more than the entire European Union 2010 targets for renewable electricity generation in Ireland, the UK, France, Germany, Denmark, the Netherlands, Belgium, Luxembourg, Spain and Portugal combined. It is also equivalent to almost thirteen times the electricity consumed in Ireland and more than the total electricity consumed in the UK. It is a unique opportunity to achieve energy independence from a natural, clean resource on the nation's doorstep.

It takes time to establish the means to harvest wind energy. Sites must be carefully selected; cost competitiveness proven; funds raised; planning permission sought and granted; infrastructure built; and the existing electricity grid adapted. The practical approach is to set goals to increase the use of wind energy relative to the current starting point. There were almost 1,300 MW of wind capacity installed in Ireland by the start of 2010. These turbines could yield some 3,800 GWh (or 3.8 billion kWh) per year, working at a third of their capacity. The Irish government has set the target of producing 40 per cent of Ireland's electricity from renewable sources by 2020, with the bulk of this renewable electricity coming from wind. Achieving a slightly more ambitious target of 50 per cent in line with the enormous resource available would require roughly an additional 170 offshore 6 MW turbines and 1,350 onshore 3 MW turbines, all operating at a third of their capacity. The onshore turbines would be situated over some 350 square km of land or 0.5 per cent of Ireland's land mass, allowing for an optimal spacing of just under 0.25 square km per turbine. The actual footprint of turbines is less than 5 per cent of this spacing required. Farm work can continue and animals can graze unaffected in their midst.

Is the amount of additional turbines required, 1,520, a large number? Some comparisons can be helpful to set the number in perspective. It is significantly less than the number of mobile phone transmission 'live sites' that have been erected in Ireland over the fifteen years to 2010, of which there are over 5,800; it is less than the number of church steeples, of which there are some 2,000; it is much less than the 18,400 medium to large electricity voltage transformers on the ground and the 212,000 thousand pole-mounted electricity transformers. With generous spacing, the

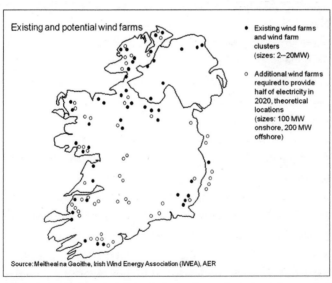

Achieving relative electricity independence in Ireland from wind energy

turbines would take up a land area that is roughly half the combined area of the sports pitches of Ireland's 2,300 GAA clubs and 4,000 state supported schools. Do we see these other amenities? Yes. Do they impinge on our daily lives? Not really. We accept them in the landscape we have shaped in order to lead socially meaningful lives. An illustrative map of the amount of wind farms required to provide relative electricity independence is shown on the previous page.

The growing wind energy sector in Ireland has given rise to significant investment, company development and employment. There were 123 wind farms in place by the start of 2010. Key owners and operators of existing wind farms in Ireland include private companies such as Airtricity, which began with support from NTR (formerly known as National Toll Roads) and has invested nearly €1 billion in wind farms in Ireland. Airtricity is building a Europe-wide portfolio of renewable energy facilities with its current parent company, Scottish and Southern Energy. Other companies include Gaelectric, Eco Wind Power and SWS Group as well as state-related energy providers Bord Gáis, Bord na Móna and the ESB, which is devoting some €11 billion to sustainable and renewable energy projects in the decade up to 2020.

Irish companies can also play a role in the global wind energy sector. Several Irish companies such as Mainstream Renewable Power are developing wind farms across the US, Latin America, Europe and Africa. Furthermore, high-tech applications are being developed by Irish start-ups that may shape the way wind is harnessed and managed. For instance,

Servusnet, a start-up launched by four former Motorola executives, has developed a software package that helps improve monitoring and performance of wind farms, cutting costs and maximising revenue. The technology is an example of an Irish-based development that can be exported and applied to any wind farm in any corner of the world.

In Ireland, 11,000 jobs could be created from the work required to install and operate the additional wind turbines required to take full advantage of the country's wind resource. The capital cost of the turbines would be about €5.5 billion and a further €5 billion would be required for commissioning and local services. These costs would be spread over ten years. The cost to achieve a major step towards national energy independence is less than other national investments over shorter periods. For instance, the total cost is similar to the amount committed to support the country's three largest banks in an initial and immediate rescue investment in early 2009, roughly a third of the total of state and taxpayers' funds singularly required to bail out the worst affected bank, Anglo Irish, as detailed in June 2010 and less than a quarter of the investment needed to shore up all Irish banks and building societies, as detailed in March 2010. The variable costs of electricity produced subsequent to capital investment would effectively be zero because wind is free, though nominal administrative and maintenance costs would be required. A positive return on investment would be achieved from the sale of the electricity to customers over the lifetime of the turbines. This wind energy would also result in a yearly reduction of €200 million of imported fossil fuel, annual savings of €50 million in reduced carbon costs and an annual contribution of €50 million in taxes and other benefits to the economy.

When we consider that wind energy is best complemented by other sources of energy in order to manage the intermittent nature of its supply, then a 50 per cent target is a reasonable medium-term goal for Ireland. However, this should not prevent Ireland from producing additional clean wind electricity and exporting it to the UK and mainland Europe.

Denmark has successfully demonstrated how wind energy can contribute significantly to energy freedom. Despite having a mere fifth of the wind resource available in Ireland, wind energy already provides more than 20 per cent of Denmark's electricity use. Denmark has installed over 5,200 turbines, employs more than 28,000 people in the industry and plans to double wind energy capacity by 2020. Lessons can also be learned from the Danish experience. A reliance on interconnectors to Norway and Sweden that enable both the export of surplus wind energy and access to electricity supply when the wind does not blow underlines the importance of additional interconnector capacity between Ireland and the UK and countries in mainland Europe such as France. Options for storing wind-generated electricity must also be explored. Adopting a long-term view on rising fossil fuel prices has enabled the Danes to establish a major defence against the increasing cost of imported fuel by accepting some short-term price increases. The Danish precedent demonstrates what is possible. Ireland can apply both the successes and lessons.

It is possible is to store the vast amounts of wind energy that are available in Ireland by converting it to the potential energy of water pumped to a higher elevation, compressed air or hydrogen. This would overcome the 'stop-go' limitations of wind by introducing a consistency that more closely matches the nature of our demand and the power quality required by our legacy electricity grid. For instance, the 'Spirit of Ireland' initiative proposes using wind energy to pump sea water into natural coastal valleys, where it can later be used to generate hydroelectricity.

There are four requirements for accelerating the introduction of wind energy in Ireland. The first is to continue to increase the level of open competition and move more quickly towards liberalisation of the electricity market. The second is to facilitate the upgrade of the electricity distribution and transmission networks so that they can accept wind capacity and load. The most windy and valuable locations may also be remote from current electricity grid infrastructure. The third is to improve the planning process by reducing costs and waiting times. Lastly, collaboration by government, multilateral lending institutes such as the European Investment Bank and commercial banks to support long-term loan requirements for wind farms is needed to enable equity providers and entrepreneurs to move faster to build Ireland's clean energy supply.

If Ireland were to supply half of its electricity from wind by 2020 and continue to harness wind energy for local use and international export thereafter, it would stand out as a world beacon for how to capture clean energy and stride towards energy independence.

SUN

> 'All bright and glittering in the smokeless air.'
> - William Wordsworth,
> *Composed Upon Westminster Bridge.*

The importance of sun as the giver of life cannot be overstated. Without it, the earth would be cold and dead. The sun is the source of all main energy forms on earth, while geothermal energy percolates from elements once created in the sun or another similar star and nuclear energy is a faint echo of processes that have been occurring in the sun and universe for billions of years.

BIRTH AND DEATH OF A STAR LIKE OUR SUN

We often think of the space between stars as a black vacuum. However, gas and dust are found in large amounts throughout this space. The gas is comprised mainly of hydrogen and helium but also water, carbon monoxide, ammonia and even alcohol. The dust includes graphite and silicate, which is like sand. Dust particles are about the size of those found in cigarette smoke. Although the gas and dust are spread very diffusely through space, there are vast amounts because of the vastness of space – enough to provide the material for star formation.

Gas and dust can settle together in the coldest places of the universe to form enormous interstellar clouds that can be over 150,000 times as wide and long as our own solar system. A low temperature of about -250 °C results in very little pressure between the particles and gravity begins to draw them together. This collapse builds a core mass, which increases the strength of gravity and makes particles fall faster and generate heat. The core of this 'protostar' generates enough heat and pressure to emit radiation that can blow away outer parts of the clouds, which may condense to form planets. The protostar stage lasted for some 30 million years in the case of our sun. Eventually, the inner core of the protostar becomes so hot and dense that nuclear fusion of hydrogen begins and a mature star is born.

The energy and pressure released from hydrogen fusion counterbalance the collapse of gas into the core, creating a steady state. If the rate of fusion were to increase above its steady state, the temperature would rise and gases would expand away, reducing the level of fusion. In equal balance, if the rate of fusion were to decrease below its steady state, the temperature drop would result in contraction and more fusion would occur. Our sun has experienced this steady state balance for about 4.5 billion years.

The overall balance belies the turmoil of the fireball, with a surface temperature of 5,500 °C and 15 million °C at its core. Twisting magnetic fields, enormous explosions of sound and an immense wind of ejected particles lacerate the broiling, churning shell of the sun. Molten flares erupt from the surface and shoot up to a million kilometres high. Flares and solar wind can blast energy and particles as far as the earth's atmosphere, where they create hauntingly beautiful auroras seen near the poles, knock out satellites, cause power surges on electric lines and induce static in radio signals.

There is enough hydrogen in the sun's core to fuel this energy furnace for another 5 billion years. At that time, the steady state will start to falter and the sun will initially contract. This increased density will actually lead to a jump in temperature that sets off fusion of the remaining hydrogen at the outer shell of the sun, swelling it into a red giant. Our sun will be 1,000 times brighter than it is today and the earth's atmosphere will be swept away by an intense solar wind. Helium 'ashes' from fusion will be dumped into the core making it denser and hotter until it ignites in a violent flash of helium fusion. This new source of helium nuclear energy will extend the sun's life for a final 100 million years. It will expand in size to swallow Mercury, Venus and eventually Earth and will be 10,000 times brighter than it is today. As the giant approaches our planet, the earth's solid crust will melt and vaporise and the residual mass will slow its rotation and become dragged into the sun to become one with it.

When the last of the helium is spent, a solar wind travelling at 4 million km per hour will blow away the outer layers of the sun and a hot core of dense carbon will remain as a 'white dwarf'. The core will cool over several billion years to become a burned-out, degenerate cinder, hurtling through cold space.

SUN'S PLACE IN THE UNIVERSE

The sun is a tiny speck in the infinity of the universe. To an earth observer it seems large, with a diameter of 1.4 million km, 110 times that of the earth. The sphere of the sun could accommodate a million earths. However, many other stars are known to have masses up to 150 times greater than the sun. The size of the largest known star, VY Canis Majoris, is estimated to be 2,100 times that of the sun. The sun resides in a swirling galaxy that contains about a further 200 billion stars. This Milky Way galaxy spans a distance of 160,000 light years or 1.5 trillion km, more than 10 billion times the distance between the sun and the earth. It is estimated that there are more than a further 100 billion galaxies and 70 billion, trillion stars within the visible range of modern telescopes. The actual number of stars beyond our vision could be infinite.

SUN IN ANCIENT CULTURE

The sun has had enormous cultural and religious significance throughout the history of humans. The pantheon of international sun gods includes the Irish Celtic Lú, the Egyptian Ra, the Chinese Tai Yang Shen, the Greek Helios, the Roman Sol, the Hindu Surya or Mithra and the Incan Apu Inti.

Most cultures celebrated the seasons and daily movements of the sun that were seen as reflections of life's rhythmic cycles. In Ireland, bonfires were lit at the summer solstice and festivals held in the month of August (Mí Lúnasa in Irish), which was dedicated to the sun. In Roman times, the winter solstice was marked over seven days by lighting candles, decorating homes with holly and giving presents. The winter solstice of 25 December in the Julian calendar was the date chosen by Pope Julius I in the fourth century as the celebration of Christmas in an effort to replace pagan sun worship. In China, the daily rising of the sun was greeted by sprinkling tea leaves and evening ceremonies marked its setting. In the Aztec and Incan cultures, humans and animals were offered to appease the sun god at its summer solstice, by cutting the pumping hearts from living sacrifices.

These cultures and all humans have recognised the power of the sun on life since the first known dawn. In our lifetime, that power can be captured on a massive scale for the first time to provide clean energy.

CAPTURING SOLAR ENERGY: HOW DOES IT WORK?

Energy emitted from the sun results from multiple fusions of four hydrogen nuclei forming one helium nucleus. Some 626 billion kg of hydrogen are converted to about 622 billion kg of helium every second. The difference in mass between the start and end elements is released as energy, primarily in the forms of light and heat. 'Bundles' or 'photons' of sunlight travel in a wave-like fashion through space towards earth. It takes eight minutes for light to travel the 150 million km to our planet. When it arrives, the light energy is partly absorbed by the objects it strikes. The absorbed energy causes the atoms and molecules of objects to move faster. This movement causes friction and is measured as a rise in temperature. In this way, light energy is converted to local heat energy.

White light emitted by the sun is made up of a spectrum of energies, all moving at different wavelengths. These can be seen when white light is separated in a prism or a rainbow. A white object absorbs none of these energy wavelengths and all are reflected, so full white light is registered by our eyes. A black object absorbs all of these energy wavelengths and reflects none, so no colour, or black, is registered by our eyes. An object of a particular colour absorbs every energy wavelength *except* the energy with a wavelength of its own colour, which it reflects and is registered in our eyes as the colour of that object. The different wavelengths have different energy intensities. Solar energy collectors are designed to absorb those wavelengths with the highest levels of energy intensity.

There are three key approaches to capturing solar energy. The simplest approach is to capture the heat energy that results when sunlight is absorbed by an object. A 'solar thermal collector' contains a liquid that easily absorbs sunlight. The liquid is present in pipes or plates that catch the sun's rays throughout the day. The liquid can reach temperatures of up to 100 °C and is circulated from the pipes to a reservoir from which the heat can be transferred to water or air. This heated water or air can then be pumped to locations where the heat is needed, such as a shower, swimming pool or air vent. The efficiency of converting solar energy to electrical energy in this way can be as high as 50 per cent.

A second approach of 'concentrating solar power' can be applied to electricity generation or industrial heat processes. Sunlight is concentrated on one spot by either reflection from multiple mirrors or a magnifying lens. The mirrors and lenses can be programmed to track the movement of the sun throughout the day so that they constantly reflect the most powerful sunlight. Temperatures of the concentrated sunlight can reach 10,000 °C. This heat can create steam to drive turbines for electricity generation. The efficiency of converting solar energy to electrical energy in this way is up to 25 per cent. The concentrated heat can also fuel an industrial furnace or be stored in molten salts for the subsequent release of heat when the sun is not shining.

A third approach to capturing the sun's energy is to expose plates of crystals in which electrons are knocked loose by sunlight energy to create an electric current. These 'photovoltaic' crystals, often silicon, readily absorb light energy that is just large enough to loosen electrons and raise them from a state of low energy to a state of high energy. The photovoltaic cells contain two sections of silicon crystal with slightly different properties and a junction where they meet. When the sunlight strikes the crystals, electrons are energised and move about freely only in one silicon section. Some of these free electrons gather at the junction but cannot cross it and a build-up of negative charge is formed. If a conductor is attached between the two sections, a current of electrons will flow across it, similar to the current that flows between two poles of a battery. The cells can be connected in an extended sequence to create a large potential for electricity.

Photovoltaic panels are robust, require little maintenance, do not create any emissions in operation and work in silence. Panels can be arranged in arrays in large-scale ground installations to produce electricity. 'Building-integrated photovoltaic' panels can be incorporated into the roofs, walls or windows of a building during construction to provide the building's energy needs. Surplus electricity can be sold into a national grid. The efficiency of photovoltaic panels in converting sunlight to useful energy is typically only some 15 per cent. The efficiency can be increased towards 40 per cent by stacking more expensive layers of photovoltaic panels on top of each other.

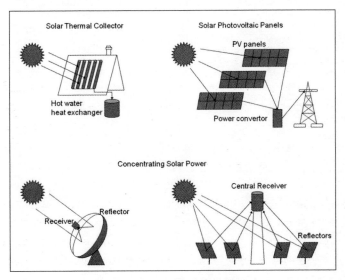

Solar energy device schematics

HOW MUCH SOLAR ENERGY IS AVAILABLE?

The total energy emitted by the sun is an enormous 3.3 million YWh (3.3 million, billion, trillion kWh) per year. The sun's energy is emitted into space in all directions and the earth intercepts just 50 billionths of this. The energy that reaches the planet is equivalent to 100 million thunderstorms. The fraction that causes actual storms on earth results in an average of 100 bolts of lightning striking the earth every single second.

About 40 per cent of solar energy is reflected away or absorbed by clouds and air particles and some 60 per cent of the energy strikes the ground. The solar energy reaching the earth's surface is some 950 EWh (or 950 million, billion kWh) per year, which is 6,900 times the total world energy

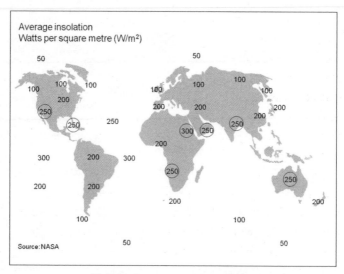

Global solar energy rates available

consumption at present. Put another way, an hour and a half of solar energy reaching the earth would be more than sufficient to power the world's energy needs for a full year.

The highest rate of sun's energy radiation, known as insolation, that strikes a square metre of ground is about 1,000 W (1,000 W/m^2), which would yield 1,000 Wh of energy in one hour. Insolation varies with the time of day and season and the map above shows the variation of average insolation with geography.

The world's highest insolation is found in the Red Sea and is 300 W/m^2 on average during a year. Average insolation across the world is 170 W/m^2. Collected over a year, this is roughly equivalent to the energy in a barrel of oil in every square metre.

We would need to capture just 0.0145 per cent of the solar energy reaching the earth's surface to meet the world's total current energy needs. If we were to use solar energy collectors that converted solar energy with just 15 per cent efficiency, they would cover 0.1 per cent of the earth's surface or just under 0.3 per cent of land surface. The source of this energy will not run out for the lifetime of the earth and its daily collection produces no greenhouse gas emissions.

Solar energy is not distributed evenly across the world. The areas of highest insolation are concentrated within a band of 3,000 km south and north of the equator. However, the areas of the world that use energy most intensively tend to be concentrated in industrialised areas 4,000 to 5,000 km from the equator. While solar energy can be captured in every location with some effectiveness, some industrialised countries could continue to be reliant on foreign energy supplies if solar were the only primary alternative. A tension between areas of energy source and energy use would result, similar to the tension that exists due to the uneven distribution of hydrocarbon resources. Furthermore, a means to transport energy long distances would need to be established. Hydrogen generated from solar electricity could serve as a convenient means of energy transport and could be transported by pipeline or in liquid form by sea.

Twelve European companies have proposed the location of concentrating solar power systems and photovoltaic panels across 17,000 square km of Saharan desert with a view to meeting up to 15 per cent of Europe's electricity demand. Key participants include Deutsche Bank, Siemens, ABB and Abengoa, under the auspices of the Desertec Foundation.

The cost of electricity generation from solar energy at 15 per cent efficiency is still more than four or more times the average cost of generation from fossil fuel sources. Despite this, solar energy has found a solid foothold as a result of easy installation, ability to supply energy at expensive peak hours and government incentives.

Worldwide solar energy capacity stands at just 0.3 per cent of global electricity capacity. However, the sector enjoys the highest growth rate of all alternatives, increasing rapidly in size by 40 per cent each year.

Japanese, German, Chinese and US manufacturers are leading improvements in cost and efficiency and these factors will certainly drive continued growth in use of solar energy. Achieving these improvements is not a fundamental obstacle but an engineering hurdle that can and will be surpassed.

One of the greatest masters of engineering, Thomas Edison, confided his belief in solar energy during a conversation with friends Henry Ford and Harvey Firestone in 1931. With only months to live, the man who conceived the incandescent light bulb and a range of other electricity-thirsty inventions said: 'We are like tenant farmers chopping down the fence around our house for fuel when we should be using Nature's inexhaustible sources of energy – sun, wind and tide. I'd put my money on the sun and solar energy. What a source of power! I hope we don't have to wait until oil and coal run out before we tackle that. I wish I had more time.'

SOLAR ENERGY IN IRELAND

Ireland is not known for an abundance of blazing sunshine throughout the year. Millions of Irish and international sun-seekers do not flock to sprawling seaside resorts here to bathe on sun-drenched beaches from May to September. Conversations among people in Ireland often feature such phrases as: 'what summer?', 'bring an umbrella' and 'we can't complain, we had a few sunny days in June'. What then, are the chances that Ireland can achieve any energy independence from the sun that so often seems to desert the country?

Solar energy potential in ireland

Ireland has an insolation level of 105 W/m^2 on average during the year. This is about three times less than the highest insolation level in the Red Sea, less than half of that in Spain or Arizona and is about 60 per cent of the world average of 170 W/m^2. However, the fact remains that the sun radiates energy at an average rate of 105 W/m^2 that could be tapped. Other countries provide inspiration. Germany has just a modest average insolation of 145 W/m^2 yet has managed to generate more power from the sun than all the rest of the world put together. The Netherlands experiences an insolation level just slightly higher than Ireland's but produces more power from the sun than either France or Italy, where the sun is plentiful.

The total theoretical solar energy striking Ireland's land area is some 65 PWh (or 65 trillion kWh) per year, which is about 2,200 times the current electricity consumption in Ireland. It would be naturally impossible to capture all of this energy efficiently. Despite the Irish climate, both solar thermal collectors for water heating and photovoltaic panels for local electricity generation can operate successfully, though concentrating solar power systems are not effective. If just 0.5 per cent of land was covered in solar thermal collectors and another 0.25 per cent of land in photovoltaic panels, then about 185,000 GWh (or 185 million kWh) could be supplied each year, taking conversion efficiencies into account. This is almost equivalent to the total current energy consumption in Ireland.

Expense remains a serious barrier to harnessing solar energy in Ireland as it costs some four times more than conventional energy sources. Furthermore, some might argue that the concept of comparative advantage suggests Ireland should leave large-scale solar energy generation to those countries blessed with high insolation levels and focus instead on natural resources that are advantaged in Ireland, such as wind and ocean. However, the significant amount of solar energy that is theoretically available in Ireland, falling technology costs, improvements in technology for low insolation levels and rising fossil fuel prices all underline the potential of solar energy for Ireland in the long term.

The sun can still play a helpful role in contributing to energy independence in the short term. If just one fifth of households placed 10 m^2 solar thermal collectors on their roofs and if energy-conscious companies used photovoltaic panels that replaced just 5 per cent of electricity by 2020, some 3,000 GWh (or 3,000 million kWh) could be supplied each year. This would be roughly equivalent to 3.5 per cent of Irish heat and electricity consumption.

Ireland has already installed solar thermal collectors that produce some 250 GWh per year and compares well to other European countries. Ireland is fifth in Europe for the installation of solar thermal capacity on a per capita basis, behind only Cyprus, Austria, Greece and Germany. This has been facilitated by support for solar thermal collectors from the government such as grants under the 'greener homes scheme' managed by the Sustainable Energy Authority of Ireland. However, Ireland is well below average for the installation of photovoltaic systems and has ground to make up here.

The adopters of solar energy technology are leaders who value the benefits of emitting less carbon and are comfortable to wait several years before the savings from free solar energy make up for the cost of installation. The economic feasibility of solar energy in Ireland is on a constant path of improvement. In the very near future, solar electricity will be generated from thin film photovoltaic transparent glass that will cost the same as double-glazed windows and from paint that can be pasted on the outside of any building. Falling prices and rising efficiencies will continue to make solar energy more appealing to a growing number of early adopters.

Ireland can produce companies that compete on a global basis in the solar industry. For example, NTR plc, an international renewable energy group, is pioneering the design and development of concentrating solar power systems for international markets with investments in Stirling Energy Systems and sister company Tessera Solar. Their 'SunCatcher' technology comprises a mirrored concentrating dish and an efficient engine that converts the sun's energy into electricity. Some 70,000 SunCatchers brought into operation in Arizona and California to provide 1,750 MW of contracted capacity will be just the beginning of a planned worldwide roll-out of the systems.

Ireland's place in the world of solar energy can also be as a leader in developing and adopting technology that fits our climate, setting the standard for large industrial areas subject to cloudy conditions in northern America, Europe and Asia.

For example, Irish university researchers are leading global development of cheaper materials that are more suitable for capturing the energy from diffuse sunlight prevalent in cloudy locations. Technology jointly developed by three universities imitates the structure of green plants that can efficiently absorb the sun's energy even when its rays are obscured. Solarprint is an Irish company that is developing 'third generation' photovoltaic technology called 'dye-sensitised solar cells'. This technology achieves breakthrough improvements compared to conventional photovoltaic panels as it produces power in cloudy conditions, at all angles of incidence to the sun and even from artificial indoor light. The Solarprint cells can also be printed on to surfaces, reducing manufacturing cost and carbon footprint. The fact that the world's largest industrialised areas with the greatest demand for energy are located in relatively cloudy latitudes underlines the importance and opportunity for such breakthrough technology development in Ireland.

OCEAN

'The wild unrest, the snowy, curling caps,
That inbound urge and urge of waves,
Seeking the shores forever.'
— Walt Whitman, *From Montauk Point.*

MAKING WAVES

Waves resonate with a sense of life within us. They captivate us with their creativity, uniqueness and the glory of their last act. The energy of waves is powerful and unforgiving. It can quench a life in seconds. It can twist the metal and smash the timber of craft and erode a rocky coast. Yet it speaks to us of childhood. It entices us into its folds for rejuvenation. Gathering up this relentless energy can give us freedom.

Waves are the means of transporting energy through any material. A consistent source of energy will create waves, or oscillating pulses, through steel, water or air. When energy passes through a medium in a wave, it creates an efficient passage by reordering the material in its way. When energy passes through water, for instance, the water particles undergo a circular motion, being drawn towards the wave, lifted to its crest, dropped to its trough and then regaining equilibrium. Water effectively remains in the same place in the wake of a wave of energy that passes through it. Vast quantities of energy pulse through oceans that cover two thirds of the earth's surface. Although waves captivate us with their ever-changing forms, their energy is predictable and therefore can be captured and put to work.

The four key forces that create waves in the oceans are: wind; seismic movements; the gravity of the earth, moon, sun and other planets; and the rotation of the earth. Most conspicuous among these is wind, which transfers its energy to water as it blows over it. Wind buffeting the surface of water will create ripples. The surface tension of water and the earth's gravity will work to rebalance the water's surface. However, if the wind is consistent, it will create choppiness and continue to increase the size and strength of waves. Eventually a regular swell will be formed from the toppling disorder of chop and will move in the direction of the original wind, even after the wind has ceased. The strength of the wind, the duration of its blowing and the amount of open water on which it has acted will determine the size of the waves. Waves will continue in a particular direction until they meet an opposing action, such as another wind or landfall. As the depth of water shallows beneath the wave, the crest will rise up and finally crumple under its own weight and crash upon the shore in a final gasp of individuality.

Seismic tremors, earthquakes and volcanoes can create a shock impact on the pressure of a body of water. They can raise a wave smaller than 1 m in the deep ocean but with such intense energy that it can travel at up to 800 km per hour with a gradual slope across a wavelength of 200 km. As this tsunami encounters shallower water, it slows down and its massive energy is transferred from speed to height. It can devastate the coastline in its path and drag the remnants back to sea in retreat. The Indian Ocean tsunami of 26 December 2004 claimed some 230,000 lives and displaced 1.7 million people following a seabed earthquake that released 1,500 times the energy of the Hiroshima bomb on the surface of the earth.

Tides are the world's longest waves and span half the circumference of the globe. The gravitational energy of the moon primarily, though also the sun and other planets, combine to pull the fluid body of the sea towards it as the earth spins beneath it. This creates a bulge of water that moves at some 1,200 km per hour. The high-tide crests and low-tide troughs of tides pass each point on earth in a period of nearly twelve and a half hours.

All of these forces combine to create waves that feature on 70 per cent of the earth's surface. The randomness of their surf-beat belies the orderliness that physicists and oceanographers have applied to them. Observers have gleaned enough from the patterns that are framed by the laws of physics to know how to measure wave energy and begin to net it for our benefit.

CAPTURING OCEAN ENERGY: HOW DOES IT WORK?

Wave energy

Waves have maximum energy in deep-sea locations, where water depth is over 40 m and energy has not been dissipated by a shoaling seabed. There are several approaches to converting the deep-sea kinetic energy of waves into electrical energy. For instance, the bobbing motion of a floating device can create a pumping action that forces water through a turbine and generates electricity (e.g., Wavebob). Alternatively, connected floats that are lifted and dropped by waves can pull and release a pressure hose that drives water through a turbine for electricity generation (e.g., Pelamis). A different floating device with two hinged wings that each rise and fall in sequence as a wave passes can also create a pumping action that can be put to work (e.g., McCabe Wave Pump).

Closer to shore, several different wave energy devices can be applied. A fixed cylinder that is semi-submerged, open to the water below and closed on top, allows a wave to enter

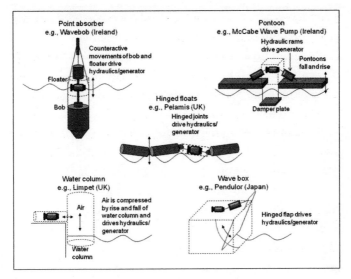

Wave energy device schematics

the device and compress the trapped air, which may be channelled to turn a turbine (e.g., Limpet). A box with a hinged flap that swings as a wave enters and leaves the box can drive a hydraulic pump (e.g., Pendulor). A further approach consists of a tapered channel that can be designed to increase the height of a shore-bound wave and the elevated wave will spill into a reservoir, from which the water is released to turn a turbine.

Tidal energy

Two key approaches to capturing tidal energy are by way of dams and turbines. A dam takes advantage of a difference in water levels across a barrage as a tide floods or ebbs. Water is released through a gate and drives an electricity generator. A successful tidal energy dam with a 240 MW capacity has been in place in Brittany since 1966. The Bay of Fundy in Canada and the Severn Estuary in the UK are sites well suited to similar tidal dams.

A tidal turbine is similar to an underwater wind turbine. Submerged turbine blades are turned by tidal currents moving at speeds of up to 15 km per hour. The blade movement is converted to electricity generation in the turbine. Tidal turbines have been successfully installed in Strangford Lough in Northern Ireland, the Orkney Islands in the UK and the Bay of Fundy in Canada. Sea water has greater density than air and the energy that can be captured per length of blade is much higher than that from wind. A 15 m diameter tidal turbine can generate as much electricity as a 60 m diameter wind turbine, in moderate tidal current speeds of 8 km per hour. Tidal turbines can be located relatively close to shores. They also have less impact on the environment than a dam system, which may be an obstruction for sea life and boat movement and result in the build-up of silt.

The ocean is an exciting frontier as the race is wide open to find a winning device for efficiently capturing the enormous amount of available energy. The prize is great and wave and tidal energy devices are under development throughout the world. The front-running developers are in Ireland, the UK, Norway, Canada and the US. The primary challenges they face are to increase conversion efficiency and

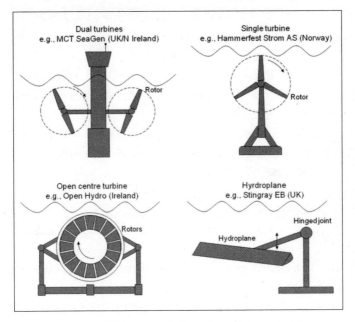

Tidal energy device schematics

reduce the cost of the devices per unit energy produced. The efficiency of converting wave and tidal energy to electrical energy is currently as low as 10 to 15 per cent, though this will rise with technology improvements, while the cost is some three to five times the cost of fossil fuel produced energy, ignoring external negative costs of fossil fuel energy. A further operational challenge is to design devices that can withstand the battering of the harsh ocean environment and minimise maintenance needs. The advantage of zero greenhouse gas emissions from daily operation, leaving installation and maintenance aside, is a major incentive for governments, multilateral agencies and traditional utilities to put their shoulder to the wheel and support inventive and entrepreneurial ocean energy companies in their efforts to harness energy from the waves. Additional significant benefits will result from successful ocean devices. For example, wave and tidal energy can also be applied to the desalinisation of sea water, which will become increasingly valuable as world freshwater supply is squeezed and demand continues to grow rapidly.

HOW MUCH OCEAN ENERGY IS AVAILABLE?

Wave energy

The total kinetic energy of surface waves is some 73 PWh (or 73 trillion kWh) per year throughout the planet's oceans. This is half the world's total energy consumption and over four times the world's electricity consumption at present.

The energy of a wave in deep water is proportional to the time interval between waves (the period) and the square of the height of the wave. Therefore, a wave in a moderate Atlantic swell, 3 m high with a period of 8 seconds, will carry 36,000 W per metre of wave front (W/m). This is the same amount of energy that could power 600 light bulbs each rated at 60 W. In a storm, a 12 m high wave with a period of 15 seconds will carry over 1 million W/m, equivalent to the rate of energy consumed by 1,500 households.

The amount of wave energy that could practically be exploited is estimated to be just some 2 PWh (or 2 trillion kWh) per year, equivalent to just over 11 per cent of world

Early oil rig blowout

Military-protected
flaring oil well in
the Middle East
(courtesy of PA)

Glacier under threat from warmer climate in Argentina (courtesy of JT)

Solar flare and the solar surface, churning at 5,500 degrees Celsius (courtesy of NASA)

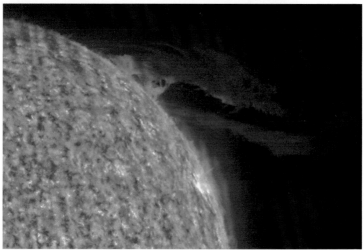

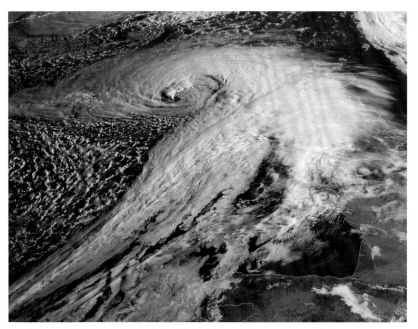

Storm winds approaching Ireland, which experiences some of the strongest winds in the world (courtesy of EUmetsat)

Wind power sculpting a tree (courtesy of Steve Barnett)

Gabe Davies surfing the Aileens wave off County Clare. Ireland experiences some of the most powerful waves in the world (courtesy of Waveriders)

Sugarcane grows prolifically in the tropics

Kings Mountain wind farm, County Sligo (courtesy of IWEA)

SunCatcher™ technology in Arizona, manufactured and deployed with investment from Irish company NTR (courtesy of NTR)

Third-generation solar technology for cloudy climates developed by Irish company, Solarprint (courtesy of Solarprint)

Wavebob harnessing ocean energy in Galway Bay (courtesy of Wavebob)

OpenHydro tidal energy turbine and test rig (courtesy of OpenHydro)

Miscanthus biomass thriving in County Meath (courtesy of Tom Bruton)

Oilseed before the harvest in County Meath (courtesy of Tom Bruton)

Sellafield nuclear reprocessing plant, Wales (courtesy of Simon Ledingham)

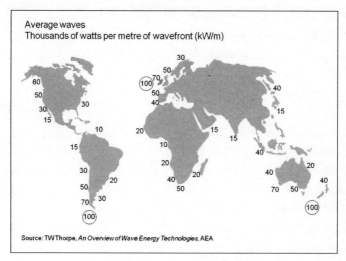

Global wave energy rates available

electricity consumption. However, a further key constraint for feasible wave energy is the ability to compete on cost with existing sources of energy. Wave technology has yet to capture the energy of the waves at a cost that is lower than conventional sources. Until technology costs are lowered or the costs of other energy sources rise sufficiently, wave energy will remain in development. The technology is on a path of constant improvement and optimistic experts estimate that some 5 per cent of global electricity could be derived from wave energy by 2020.

Tidal energy

The energy of tides that surge throughout all the oceans is more than 3,500 PWh (or 3,500 trillion kWh) per year, which is twenty-five times the world's total energy consumption and 200 times the world's electricity consumption at present. This enormous amount of energy is particularly difficult to harness and its potential collection is limited to a few advantaged locations where tides rush through narrow channels or the difference between high- and low-water marks is large. A key advantage of tidal energy is its predictability and therefore its ease of absorption into an existing national electricity grid system. As with wave energy, further conversion efficiency and cost reductions are needed to make tidal energy feasible. Exploitable global tidal energy is considered to be only just over 1 PWh (or 1 trillion kWh) per year, equivalent to about 6 per cent of global electricity usage.

OCEAN ENERGY IN IRELAND

Waves crash against the 5,631 km of Ireland's coastline in a constant and renewable cycle. They have shaped the island's history and way of life. Over 10,300 boats are recorded to have been lost to Irish waves. Every stretch of seashore has witnessed countless life-changing episodes that echo stories from around the country's entire coastline.

In the remote southwestern corner of Ireland, Bere Island faces the Atlantic in the world's second deepest natural harbour after Sydney. Waves have brought bounty and tragedy here year after year. In 1908, high waves drove the *Bonnie Lass* from Southampton, carrying a full hold of fresh lobster, on to rocks at the east end of the island. The crew scrambled to safety while the hole in the wooden hull provided an escape for the catch and a bumper season for local lobster fishermen. In 1986, waves battered the island during a gale force storm and smashed a Spanish fishing trawler, the *Contessa Viv*, against cliffs at the west end. Local residents scrambled over jagged rocks to search for survivors while the island's ferryman and local boatmen reached into the seething waves in the middle of the channel and hauled out both breathing and lifeless men thrown from the sunken trawler. The five lives lost that night are just some of the tens of thousands that Irish waves have claimed.

Monster waves as high as 15 m on the Irish west coast draw the world's best surfers to ride the new frontier of world surfing. Surfers have described Irish waves as 'gnarly, vicious, hollow slabs' in a 'cold paradise'. Beaches, reefs and point breaks provide a range of waves for all styles and have attracted the surfing World Masters and European Championships to Ireland. The enormous 'Aileens' wave off the Cliffs of Moher comes alive in heavy swells. It can only be reached by jet ski tow-in on big wave days and is fabled as one of the planet's most beautiful and inspiring waves. Irish-Hawaiian, George Freeth, who introduced surfing to the US, would be happy with the full circle of world surfing that brings leading surfers to this country to score rides.

OCEAN ENERGY POTENTIAL IN IRELAND

Wave energy

Ireland has some of the best wave resources in the world. This is due to strong prevailing winds and the wide open fetch of the Atlantic, where waves have the chance to build up large energy. The energy of waves surrounding the island throughout 220 million acres of marine resource can be measured accurately. Seasonality is taken into account when predicting averages because, for instance, the average energy of waves in the winter is seven times more powerful than the summer average. An overall average 430,000 GWh (or 430 billion kWh) of available wave energy could be harnessed each year along a notional ellipse of 1,400 km surrounding the island, assuming an average wave energy rate of 35,000 W/m. This is eleven times the electricity consumed on the entire island.

Considering only the powerful waves that approach Ireland from the Atlantic, the wave energy crossing a straight line of 600 km is 210,000 GWh (or 210 billion kWh) per year, assuming an average wave energy rate of 40,000 W/m. This is almost seven times the electricity consumed in Ireland. The actual wave energy is likely to be higher as larger annual average wave energy rates of up to 77,000 W/m have been measured off the west coast of Ireland, among the highest measurements in the world.

Several constraints are considered when assessing the practical potential of wave energy. For instance, it is not practical to place wave energy devices along a 600 km stretch of sea in the medium term. Furthermore, the 'wave to wire' efficiency of devices, converting raw wave energy to useful energy, is currently just 15 per cent, though this is likely to rise in the future well above 20 per cent. Therefore, the average energy that is practically available is some

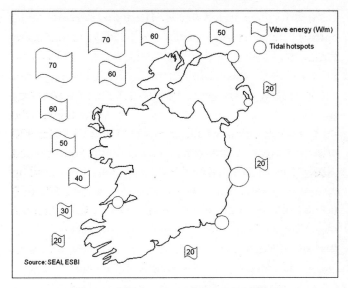

Wave and tidal energy resources in Ireland

32,000 GWh (or 32 billion kWh) per year. This is roughly equivalent to Ireland's current electricity consumption.

Tidal energy

The total tidal energy around Ireland is estimated to be some 230,000 GWh (or 230 billion kWh) per year. This is equivalent to more than seven times current electricity consumption in Ireland. However, there are only a few coastal locations with swift movement of currents where tidal energy could feasibly be captured. The practical tidal energy available in Ireland is a mere 2,600 GWh (or 2.6 billion kWh) per year, which would meet some 6 per cent of electricity consumption needs. The locations where tidal currents are strongest include the Codling and Arklow Banks, off Wicklow; Tuskar Rock and Carnsore, off Wexford; the Shannon Estuary; Inishtrahull Sound, off Donegal and the northeast coast; and Strangford Lough in Northern Ireland.

A medium-term government target is to install 500 MW of energy capacity from both waves and tides in Ireland by 2020. This capacity could produce about 650 GWh of energy per year if the devices were to operate at a productivity of just 15 per cent, edging towards 1,000 GWh if higher efficiencies can be achieved. This target acknowledges the current high, albeit declining, costs of wave and tidal energy and the pace of development of these technologies.

Achieving this target would cost about €1 billion over ten years for the wave and tidal energy devices and the associated infrastructure. The economic challenge to ocean energy is stark when one considers that building an equivalent capacity of higher productivity gas power plants would be about one third the capital cost of ocean energy. However, the argument in favour of developing ocean energy is rebalanced when the costs of fuel and carbon output over the lifetime of a power plant are compared to the effectively free and clean energy of the ocean. Furthermore, the cost of ocean energy devices will fall through economies of scale as more are produced and deployed. By way of analogy, the Ford Model T touring car was priced at $850 when it was first introduced on 1 October 1908. Just 5,986 Model T cars were sold in its first year of production. By 1916, annual production and sales of the Model T had soared to over 575,000 due to rising demand. The price of the touring car more than halved to a mere $360 mainly owing to economies of scale as the cost per unit fell with increased output.

The 2020 ocean energy target would result in a yearly reduction of €50 million of imported fossil fuel and yearly savings of €3.5 million in avoided carbon credits. More than 1,000 jobs would be created in a market valued at some €800 million. Reaching this initial target would be a first revolutionary step in unlocking the enormous potential of the ocean to help achieve energy independence.

Two of the world's most advanced wave energy devices are designed and manufactured by Irish companies, Wavebob and Ocean Energy. Prototypes have been trialled and battered by incessant waves in a test site in Galway Bay. The world's best hopes for capturing tidal energy rests on Irish devices such as Openhydro's Open Centre turbine developed in County Louth or Marine Current Turbine's SeaGen anchored in Strangford Lough.

The Irish Marine Institute has developed Galway Bay as a 'smart bay' by laying a network of buoys, seafloor cables, sensors and communication equipment for real-time monitoring to allow companies to test and demonstrate new products for the global clean energy market. The Sustainable Energy Authority of Ireland and the Marine Institute have also provided funding and research support and are developing a further world class wave-energy test site off the County Mayo coast that will feed the national electricity grid.

These government-backed initiatives and cutting-edge Irish technologies have established Ireland as a world centre of excellence for ocean energy.

Ireland and its entrepreneurs have taken a global lead in developing wave and tidal energy. There is an unprecedented opportunity for Ireland to create the winning global device for capturing ocean energy, building on this head start and the natural resource at its shores. Irish leaders are plotting a course to energy freedom not only for Ireland but for nations touched by the ocean throughout the world.

BIOMASS

'Brightness was drenching through the branches . . .
Turning silver out of dark grasses.'
– Austin Clarke, *The Lost Heifer.*

All biomass plants are effectively sun-catchers. Plants absorb sunlight energy, carbon dioxide and water and convert them to basic building blocks for growth through the process of photosynthesis. Land-based plants convert on average 1 per cent of the solar energy that strikes them into organic compounds. Water-based microalgae have the highest solar energy conversion rates among all biomass. They can convert up to 6 per cent of sunlight energy and have a theoretical maximum conversion efficiency of 12 per cent.

The energy that is captured from the sun and stored in plants can be released when a plant is eaten or burned. When plants are eaten, combustion in a living body produces biological energy for the functions of life. When burned, combustion in open air produces heat and light energy.

There are some 420,000 known species of plants in the world, with many others still waiting to be discovered. Humans have learned to cultivate some 35,000 species, of which 7,000 can be eaten and the remainder are used for flowers, gardening and landscaping. Over 5 billion people turn to plants and herbs for their sole form of medicine, while 40 per cent of the world's pharmaceutical drugs are derived from original plant extracts. The world's tropical rainforests, which cover about a tenth of the earth's land, contain over half of all known biomass species and yet only a very small proportion of these have been tested for their potential to provide medicines or life-enhancing products. About one in five of all plant species is under threat of extinction.

HUMANS AND BIOMASS

Plants – or biomass – have been a source of internal energy for the body since the beginning of humanity and have been used as an external energy fuel since the discovery of fire. Their derived heat energy enables food to be cooked and eaten and provides protection against the cold.

The cultivation of plants by humans has occurred for only a very short period in history. People have turned the soil to plant crops only in the last 10,000 of the nearly 5 million years of existence. Humans survived as hunter-gatherers for more than 99 per cent of human history on earth. Before organised agriculture, plants were gathered from where they lay or grew by providence.

The shift from exclusive hunting and gathering to organised agriculture from the earth's soil was a radical transformation. There are various possibilities as to why hunter-gatherers might have begun to sow and harvest to produce the first agricultural produce. Singular moments of discovery may have occurred to curious individuals who noted, for instance, the blossoming of a plant from a seed. Cultivated feed may have been needed for the first attempts to coax and domesticate work animals. Comparative advantage may have played a role, whereby people who were more skilled at tending plants and vines than hunting or gathering may have decided to dedicate themselves to that task in a community where the fruits of labour were shared. Growing population may have created a resource stress – a shortfall in animals to hunt and sustenance to gather, driving a need to grow food for self-sufficiency. However, agriculture occurred in places where there is no evidence of stress. This is intriguing as a greater effort is required to produce food than to hunt and gather the same amount, with the former requiring more personal energy and leaving less time for leisure. Why would one bother with agriculture in that case? Some element of human spirit must have aspired to achieve a greater good – perhaps an urge to create; to develop and expand the human limits; to shape the land on which we live; to connect with or discipline nature; to build a store in fear of an uncertain future or a hard winter; to increase personal possessions and wealth; or to feel fulfilment from productive work or the moist soil on the fingertips.

It is thought that agriculture first began in Southeast Asia, based on archaeological finds and the fact that the region was lush with vegetation, suitable soils, mild climates and fresh water when agriculture was first established some 10,000 years ago.

After some millennia of agricultural development, crops came to be seen in the classical world as divine gifts. Mediterranean cultures associated them with benign goddesses, such as Ceres in the Roman Empire, Demeter in Ancient Greece and Isis in Ancient Egypt. The Chinese attributed the health of crops to the ox-headed god Shen-nung. Aztec and Mayan mythology suggested that the god Quetzalcoatl distributed the first cereal seeds while disguised as an ant. Even the Bible places the first humans in the Garden of Eden, surrounded by plants to meet their needs. Today, the survival of humans and the cultivation of plants are still innately linked. Plants are a primary source of both food and energy and can continue to meet both needs in a sustainable way.

CAPTURING BIOMASS ENERGY: HOW DOES IT WORK?

The energy that has been absorbed from the sun and is latent in biomass can be converted into several forms of useful energy. Liquid biofuels are used for transport, while solid biomass is used for heat and power.

Liquid biofuels

The primary liquid biofuels for transport are bioethanol and biodiesel. The bioethanol production process begins with the milling of sugary crops such as sugar cane, sugar beet, grain or maize (called corn in the US). The milled feedstock is cooked at over 100 °C, which releases the simple sugars. The fungus yeast is added to the broth and it feeds on these sugars, excreting bioethanol and carbon dioxide as by-products. The bioethanol is then distilled to 99.7 per cent purity. It is a clear and highly flammable liquid in this form. Residual 'distiller's grain' can be used as a high-protein animal feed. This process is no different from the fermentation that produces bioethanol for drinking, known colloquially as generic 'alcohol', which has been practised

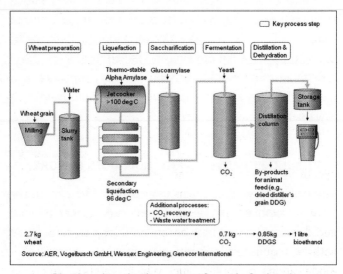

Bioethanol production process for grain feedstock

for thousands of years. The same bioethanol is distilled to around 4 per cent for beer; 12 per cent for wine; and 40 per cent for spirits. Bioethanol is produced from natural biomass and is biodegradable but at high concentrations it is also a poisonous chemical that affects the central nervous system and can cause nausea, loss of consciousness and even death if over-consumed. (Despite this, some 8 litres of pure alcohol or 125 litres of dilute alcohol are swallowed every year per person in developed societies.) At 99.7 per cent proof, it burns more efficiently and cleanly than petrol in an internal combustion car engine and provides a high octane boost. The heat energy it can produce is about two thirds that of petrol.

The biodiesel production process starts with the extraction of oils from crops such as rape seed, palm, soybean or jatropha by using either mechanical pressing or chemical solvents. The oil may also be filtered from waste cooking oil or animal tallow. A methanol catalyst is added and a process called transesterification converts the oil into fatty acid methyl ester (FAME) and a glycerine by-product. The FAME is washed and purified to create biodiesel. Glycerine may be sold for use in soap making and residual plant material can be used as a high-protein animal feed. Biodiesel burns very effectively in a diesel engine and the heat energy it can produce is almost nine tenths that of diesel fuel.

Both bioethanol and biodiesel can be blended with petrol and diesel respectively at concentrations up to 15 per cent without requiring modification to a vehicle engine. With modifications, petrol 'flexi-fuel' vehicles can burn an 85 per cent bioethanol blend, while diesel vehicles can burn up to 100 per cent biodiesel.

Biofuels are a renewable source of fuel as the crops used as a feedstock to make biofuels can be grown year after year. The energy balance of biofuels is positive, which means that the energy required to produce biofuels is less than the energy extracted from the biomass, with the extra balance coming from the absorbed solar energy in the biomass. This is a major improvement on the energy balance for fossil fuels, which is always negative.

The relative cost of biofuels compared to the fossil fuels they substitute depends on international commodity supply and demand dynamics. Supply is influenced in particular by annual harvest yields that vary with weather conditions. International biofuel prices have typically been higher than oil prices, with the exception of the high oil prices in mid-2008. Furthermore, biofuels transported between countries may incur a customs import duty, though there is no import duty for competing oil. The European customs import duty for pure bioethanol is 19.2 cent per litre, which can be as high as half the cost to produce and ship the biofuel.

Both fossil fuels and biofuels release carbon dioxide and other greenhouse gases into the air when they are burnt in a car engine. However, biofuel crops absorb carbon dioxide when they are grown and therefore effectively contribute a lesser amount of greenhouse gases over their full life cycle compared to the fossil fuels they substitute.

The savings of carbon dioxide emitted that are shown in the following graph are typical values for biofuels grown

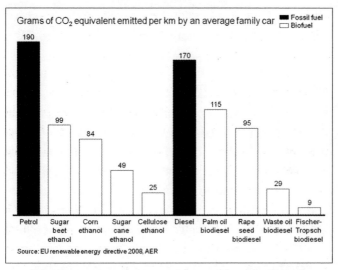

Greenhouse gas emission reductions achieved by biofuels

without changing land use in a way that would release carbon. If planting biofuel crops is accompanied by the clearance of existing thick vegetation that would otherwise absorb carbon dioxide or the draining of marshes in a way that releases methane, then the positive greenhouse gas savings of biofuels become compromised. Furthermore, intensive use of fertiliser for crop growing will have a negative effect on life-cycle emissions because of the carbon dioxide and nitrous oxide released in fertiliser production and use. Most biofuel crops are grown in a way that does not affect existing land use and does not use fertiliser intensely. Many governments have set sustainability criteria for biofuels placed on the market. In Europe for example, every litre of biofuel placed on the market must be accompanied by a guarantee that greenhouse gases emitted over its life cycle are 35 per cent less than emissions from the fossil fuel it substitutes.

Food versus fuel dilemma

A further dilemma raised by the production of liquid fuel from biomass is the possibility of competition with food. A rapid increase in global biofuel demand after 2005 led to an increase in the amount of land used to grow biofuel crops. Land and crops that might otherwise have been used for food production were diverted to biofuel production, in particular in the US. This contributed to pressure on food prices, which peaked in mid-2008. Other pressures were brought to bear on the price of food commodities, such as rising oil prices, growing population and poor food harvests because of the weather. Food prices are more sensitive to oil prices than any other variable due to the intensity of oil energy used to farm, process, package, transport and sell food. The average price of food commodities declined after oil prices passed their peak in mid-2008.

Despite the greater influence of other factors on the price of food, biofuels still contribute to the overall food price. People on the poverty line tend to suffer more from rises in food prices than those who can afford the luxury of private transport. There is no question that food for people must come first. This acute dilemma can best be solved by moving rapidly from 'first generation biofuels' to a next generation of biofuels that are produced from feedstock that do not compete with food.

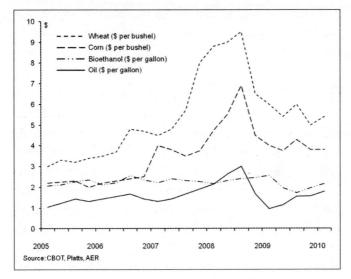

Oil as a primary driver of food price

Next-generation biofuels

The primary feedstock for biofuels that do not compete with food are cellulose and algae. Cellulose is a core part of grass and woody biomass that can be grown on land not suitable for food production. Miscanthus grass in Europe and switchgrass in the US are examples of fast-growing, high-energy cellulosic biomass. Alga is a particularly promising form of feedstock as it grows quickly, doubling in mass every day under the right conditions, and does not necessarily require arable land or fresh water. Algae can yield fifteen times more energy per unit area than first generation feedstock such as corn. Algal oils can be used in aircraft engines as well as road vehicles.

The technology required to grow sufficient amounts of these next-generation feedstock and to extract energy from them is more sophisticated than for first-generation feedstock. Hydrolysis is one approach to breaking down the complex structures of cellulose and algae into their useful energy components. In this process, either enzymes or acid are used literally to cleave and disintegrate the chemical bonds of the biomass. Pyrolysis of feedstock is an alternative approach that heats the biomass to temperatures in excess of 450 °C in the absence of oxygen. The biomass is converted to oils, gases and char, which may be refined to create energy and other co-products. Gasification is a third method for producing next-generation biofuels. Biomass is heated to above 1,000 °C in the presence of oxygen and a synthetic gas is formed. A process developed by Franz Fischer and Hans Tropsch using catalysts can convert the synthetic gas to liquid fuels, which may be used in vehicles and aircraft.

The co-products derived from biomass using these methods can also be used as high-value chemicals, pharmaceuticals and industrial materials. Such an approach to 'biorefining' is analogous to and could eventually replace oil refining, providing the full range of oil-derived products and energy from a clean and renewable source.

Solid biomass

Solid wood biomass has been used as a source of heat energy since fire was first harnessed. Wood was the primary source of energy in every pre-industrial civilisation for warmth,

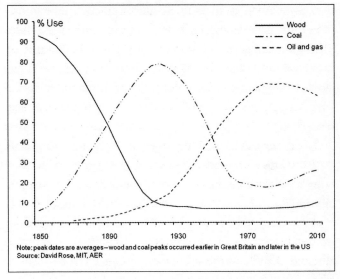

Peaks of wood, coal and oil use for energy in developed economies

cooking, light and for craftwork from bronze and gold ornamentation to horseshoe making. Wood continues to be the major source of heat energy in developing countries throughout the world. It has been displaced by fossil fuels in developed economies only in the relatively recent past.

Today, wood biomass is starting to make a comeback as the economic and environmental costs of fossil fuels that replaced it are rising and their supply becomes constrained. The key advantages of wood are that it is renewable and its combustion emits a significantly smaller amount of greenhouse gases compared to fossil fuels. Dry wood typically contains about two thirds the amount of heat energy of coal and about half the heat energy of crude oil. The chief means to convert wood biomass to energy for practical human use is to burn it. This approach might seem crude and reflect little innovation over hundreds of thousands of years of human development but it also reflects the primacy of the classical elements of nature and the unthinking efficiency with which energy seeks the easiest path for ultimate dissipation.

Wood burners come in many forms, from open hearths to enclosed stoves to modern boilers, which have regulated air-flows and heat exchangers. Simple burners have conversion efficiencies as low as 25 per cent, with the bulk of useful heat escaping to the atmosphere. Modern boilers can capture up to 90 per cent of biomass energy. The boilers are fully automated and controlled by microprocessors. A large auger feeds the combustion chamber with chip or pellets from a storage hopper at a pace determined by the energy needs of a building at any point in time. Ash falls into a collection pan that needs to be emptied only a few times each year.

Large wood boiler systems can provide the combined heat and power energy requirements of whole housing districts, university campuses, business parks and large industrial facilities. A 1 MW system could provide enough heat for 200 homes or a big hotel with a swimming pool. Systems can be as large as 500 MW and raise high-pressure steam for steam turbines connected to electricity generators. A business that replaces oil, liquid petroleum gas or electricity with wood pellets as a source of heat energy can make operating savings of some 40 per cent. When the capital cost of installing a boiler is taken into account, the time taken to pay back the

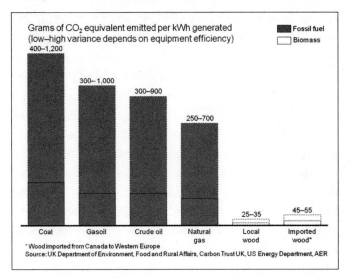

Greenhouse gas emission reductions achieved by wood biomass

investment is in the range of two to seven years, depending on the intensity of use. The lifetime of a typical wood boiler system is more than twenty years.

Burning wood does create emissions such as carbon dioxide but since the renewable growth of wood also absorbs carbon dioxide, its overall carbon footprint is significantly less than that of fossil fuels. The amount of carbon dioxide equivalent greenhouse gases emitted by wood burning is up to ten times less than fossil fuel. Furthermore, wood typically creates ten times less ash than coal and does not have the same sulphur emissions.

Wood as well as other forms of solid biomass and both animal and municipal solid waste can be converted to liquid or gaseous fuels by hydrolysis, pyrolysis, gasification and anaerobic digestion. The first three approaches are described above. Anaerobic digestion is the biological breakdown of organic matter by bacteria in the absence of air into 'biogas' and co-products. The biomass is sealed in a digester vessel and diluted to some 15 per cent. This is seeded with natural bacteria that eat into the biomass over a period of some ten to twenty days. Heating is required to the range of 30 to 60 °C and the balance of nitrogen to carbon as well as the acidity must be controlled so that the bacteria can thrive. The overall biochemical process is delicate and requires monitoring and adjustment to ensure the conditions are just right for optimal conversion efficiency, which can be as high as 75 per cent when the biomass is converted to both heat and power.

The produced biogas is comprised mainly of methane and carbon dioxide in a volume ratio of about 2:1. Biogas can be used to generate heat or power, used for household lighting and cooking, converted to transport fuel or refined to make valuable materials such as biodegradable plastic. Although the heat energy of biogas is typically 60 per cent that of natural gas, it burns more cleanly and produces less greenhouse gas emissions on a life-cycle basis.

Both incineration and anaerobic digestion make use of the latent energy in waste and avoid the long-term release of methane from waste deposited in landfill. However, incineration of municipal solid waste creates the problem of managing greenhouse gases and noxious and carcinogenic chemicals released from burning as well as the subsequent disposal of ash. Anaerobic digestion of waste that cannot be recycled also creates greenhouse gas emissions but at a lower level than incineration.

China demonstrated a successful roll-out of simple anaerobic digestion units throughout rural provinces on an enormous scale as early as the 1970s. Local communities were encouraged and supported to build small digestion units. More than 7 million units were brought into operation within five years. They consumed local renewable biomass and animal sewage to provide heat, electricity, animal feed and fertiliser, reduce emissions and achieve new-found energy independence for some 40 million people.

HOW MUCH BIOMASS ENERGY IS AVAILABLE?

The energy available from biomass is dependent firstly on the amount of sun's energy that can be absorbed and secondly on the amount of biomass that can be harvested in a sustainable way without competing with food or other necessary uses.

A theoretical maximum of just 10 to 12 per cent of the sun's energy can be absorbed and stored by biomass. The remaining solar energy is lost due to the reflection and scattering of light in the atmosphere and the fact that plants select just certain wavelengths of light to complete photo-synthesis. Plants are not perfect absorbers and in practice, the average absorption rate of land-based plants is just over 1 per cent. This solar energy fuels an average global growth rate of just under half a kilogram of biomass carbon per m^2 over a full year. The total global biomass carbon that is fixed each year is some 55 trillion kg. This consists of all grasses, shrubs, herbs, bushes, trees and any other vegetation on all fertile land throughout the world. This mass can be measured accurately on a daily basis by satellite technology. The energy of this biomass is some 460 PWh (or 460 trillion kWh), which is more than three times global primary energy consumption.

Not all of this biomass can be harvested and used for energy purposes, of course. To begin with, up to one third of it is comprised of below-ground roots that are challenging to harvest. Over a tenth resides in precious rainforests. Other tracts of land are inaccessible or environmentally sensitive. Many species do not convert to useful energy efficiently or economically. Some biomass is required for construction, paper and industrial processes. Finally, and most importantly, about 1.8 per cent of total global biomass is cultivated as food for humans.

When these factors are taken into account, a total practical and sustainable biomass energy that is available and does not compete with food or stress land use, is at least 26 PWh (or 26 trillion kWh) per year. This is about 20 per cent of global energy consumption.

In order to realise this practical and sustainable con- tribution of providing 20 per cent of our energy needs, some 7 per cent of biomass produced on the earth each year would need to be harvested and converted.

This would entail using some 8 million square km of land, an enormous area about the size of Australia but which could be distributed throughout the world.

There are 52 million square km of vacant grasslands across the world. For instance, the vast Russian and Asian steppe grasslands stretch from the Ukraine as far as Siberia. The soil is too poor for arable crops and the climate too cold for basic living. However, grasses flourish and some species grow as tall as 2 m. A small proportion of this grass could be harvested sustainably to help meet local and international energy needs.

Not all biomass for energy need be grown and harvested. The overall land area required for a 20 per cent biomass energy target could be reduced by at least a quarter by using waste generated by our own food and industrial processes. For example, the sugar cane industry generates 160 million tonnes of residual 'bagasse' waste every year after the valuable sugar has been extracted from the cane. This bagasse could provide 130 TWh (or 130 billion kWh) of energy, enough to power towns and cities throughout the tropical belt where sugar cane thrives. Several such facilities have been established in Brazil but the majority of useful bagasse is left to rot on roadsides throughout the tropics. Residual sawmill dust and forestry residues can similarly be applied to establishing clean energy independence.

Municipal solid waste, undigested organic material in animal and human waste and gas trickling from old landfill sites can all be converted to energy. Over 600 waste-to-energy plants have been built in 35 different countries but they collectively convert 170 million tonnes of waste, only a very small fraction of annual global waste.

Improving farming techniques and land management can also ease the requirement for land to fulfil food and energy needs. Farmers continue to increase the yields of crops per unit area. For example, yields of corn have increased fivefold from 2 to 10 tonnes per hectare over last 100 years.

Less than 7 PWh (or 7 trillion kWh) of energy is currently converted from biomass throughout the world every year. Most of the energy is converted to basic heat by burning wood. Only some 180 TWh (or 180 billion kWh) of electricity is generated from biomass, 1 per cent of global consumption, and about 70 billion litres of biofuels are produced around the world, under 2 per cent of world transport fuel market. Although the use of biomass for energy is growing at a firm 10 per cent per year, the total energy generated is just 5 per cent of global primary energy consumption, a quarter of what is possible using biomass in a sustainable way.

BIOMASS IN IRELAND

Ireland has among the best agricultural land in the world. Accolades of being the 'green island' are built on the confluent factors of soil, rain, gentle sunshine and a mild maritime climate, tempered by ocean currents.

Over 60 per cent of Ireland's 7 million hectares (ha) of land area is suitable for agriculture, 4.3 million ha. A quarter of the remaining non-agricultural land is covered in forestry. Even beyond the lush lands, people have sown and harvested crops for sustenance in clutches of rocky soil wherever their homestead lay. The lines of lazy beds rising up sheer mountain faces in the west are testament to the tenacity, dependence and skills of those who turned the soil.

Just a tenth of Ireland's agricultural land is used for arable crop farming, 0.4 million ha. A further tenth is used for animal rough grazing. The remainder is blanketed in grass.

Liquid biofuel energy potential in Ireland

The most practical crops that can be grown in Ireland for liquid biofuels are wheat and sugar beet for bioethanol and rapeseed for biodiesel. The total potential land area that is suitable for these crops is some 0.48 million ha. Not all

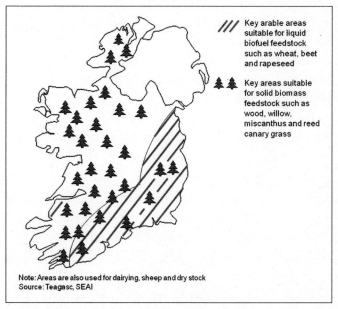

Key arable areas suitable for liquid biofuel feedstock such as wheat, beet and rapeseed

Key areas suitable for solid biomass feedstock such as wood, willow, miscanthus and reed canary grass

Note: Areas are also used for dairying, sheep and dry stock
Source: Teagasc, SEAI

Suitable land areas for growing liquid biofuel and solid biomass biofuels need be created from crops grown on valuable arable land. Recycled vegetable oil, wood residue and straw can also be used for conversion to biofuel, although the technology for wood and straw conversion continues to be developed. These sources could theoretically yield some 2 billion litres of biofuel every year, over a third of national transport fuel needs, taking biofuel's slightly lower energy content into account. For the sake of consistency in comparing across alternative energy forms, this volume of biofuel is equivalent to almost 16,000 GWh (or 16 billion kWh).

It makes little sense to compete with human food and animal feed crops to fill up vehicle fuel tanks. When arable land used for life-supporting crops is excluded, some 0.07 million ha of arable land could be devoted to energy crops for liquid biofuels. This amounts to just 1 per cent of Ireland's total land area, 1.6 per cent of good agricultural land and 17 per cent of current arable crop land. This would yield some 230 million litres of biofuel.

The yields of wheat and sugar beet per hectare of land are 8.5 tonnes and 50 tonnes respectively. However, wheat can provide 356 litres of bioethanol per tonne of feedstock, while sugar beet provides just 90 litres. Furthermore, beet is relatively more expensive than wheat to transport because of the weight of its water content and its associated earth and stones. In addition, wheat yields in Ireland are comparatively stronger than beet yields by international standards due to the attributes of Irish climate and soil. For these reasons, wheat outcompetes sugar beet as a potential feedstock for biofuel in Ireland. The former sugar factories at Carlow and

Mallow were not ideal candidates for bioethanol production facilities, despite popular sentiment and the hopes of the 3,700 sugar beet farmers, whose livelihoods were impacted with the closure of the factories after eighty years of operation. The sites would have been at a significant economic disadvantage to other Irish or international bioethanol facilities because of the capital costs required to upgrade the facilities for wheat conversion and the operating costs required to transport the finished product from inland locations to ports for fuel blending. Only a large government subsidy, prohibited by EU law, or a 60 per cent reduction in sugar beet price accepted by farmers, unlikely to be either fair or viable, could have theoretically enabled competitive bioethanol to be produced at the factories. However, the 32,000 ha of land previously devoted to sugar beet production could be applied to producing energy crops for a production facility in a more strategic location.

About 280 million litres of biofuel could be derived from the residues of the wood processing industry using next-generation biofuel conversion technology. An additional 75 million litres of fuel could be produced from waste straw. Some 300,000 tonnes of straw could be made available for this purpose from the more than 1 million tonnes of straw created each year and allowing for the amount used for animal bedding, composting and land spreading.

Recycled vegetable oil, a costly waste from the catering industry, and residual tallow from animal rendering plants in Ireland have among the highest conversion rates to biofuels, at up to 1,000 litres per tonne. Some 13,000 tonnes of recycled vegetable oil and 70,000 tonnes of tallow are created in Ireland each year. Roughly 26,000 tonnes of this waste could be made available, excluding waste used for local plant heating, unsuitable oils and allowing for losses in transport. This could be converted to 25 million litres of biodiesel.

All of these readily available sources of biofuel, which do not compete with food or stress land use, could provide some 600 million litres of biofuel each year. This is some 12 per cent of national transport fuel needs. The amount exceeds the EU 2020 target of 10 per cent renewable fuels in transport, and uses all indigenous sources. The fuel would be equivalent to just over 5,000 GWh (or 5 billion kWh).

No more than three large-scale facilities would be required to produce this volume of biofuel. They would cost a total of some €360 million of private investment to construct. A modern, medium-scale biodiesel plant has already been built in County Wexford. The use of this biofuel would avoid about 1 million tonnes of carbon dioxide emissions from the burning of fossil fuels, saving about €20 million in carbon costs annually. Dependence on the almost exclusive importation of transport fuel would be reduced, improving Ireland's import/export trade balance by some €150 million annually. Several hundred jobs would be created at the production facilities. Tax revenue of €22 million would be gathered each year. Tillage farmers would have a welcome outlet for crops, especially those affected by the cessation of sugar beet farming.

Solid biomass energy potential in Ireland

Agricultural forests cover about one tenth of Ireland: 710,000 ha. The wood in these forests is grown to maturity over a range of some twenty-five to fifty years before being felled for use in industry. Each year, these forests create several hundred thousand tonnes of waste wood that could be used for heating and power generation. The waste is created from thinning of forests, residue after final felling and milling sawdust. In total, over 500,000 tonnes of wood waste is readily available. A further 500,000 tonnes could be made available if 25,000 ha were planted each year for twenty-five years, almost doubling the amount of forestry land on former set-aside or non-agricultural quality land banks. This could provide some 2,300 GWh (or 2,300 million kWh) of annual energy immediately and a doubling of that energy supply over time. Additional energy from wood can be made available from short rotation crops such as willow, hemp and miscanthus. Some wood may also be garnered from recycling and construction waste.

The annual use of 1 million tonnes of wood biomass would reduce carbon dioxide emissions by about 900,000 tonnes and save some €13 million in carbon penalties each year.

There are several other forms of available solid biomass and some examples include municipal solid waste (MSW), meat and bone meal (MBM), landfill gas and animal manure, all of which can be successfully converted to heat and power.

Some 700,000 tonnes of MSW and 140,000 tonnes of MBM can be converted to heat and power and provide up to

2,200 GWh (or 2,200 million kWh) of energy each year. MBM has more than twice the heating energy value of peat. Gas that percolates from decomposing waste in landfill sites can be gathered and put to work, yielding some 158 GWh (or 158 million kWh) each year.

Over 37 million tonnes of animal manure is created each year in Ireland. There are more than three farm birds to every man, woman and child at any one time in the Republic of Ireland. The 14 million farm birds create 140,000 tonnes of spent litter annually. Much of this is used for mushroom composting in Counties Monaghan and Cavan. However, about 100,000 tonnes of available litter as well as spent mushroom compost could be converted to heat and power and generate some of 35 GWh (or 35 million kWh). Three times this energy is available from cattle and pig manures that are not used for nutrient spreading on land.

The total available energy that could be supplied every year from all solid biomass is just under 8,000 GWh (or 8,000 million kWh) almost immediately given current stocks of biomass resource, rising to over 10,000 GWh (or 10,000 million kWh) by 2020 with new biomass cultivation and rising again to 18,000 GWh (18,000 million kWh) in the following decade. The growth of solid biomass use is supported by government and the Sustainable Energy Authority of Ireland with the introduction of capital grants for businesses installing biomass heating and combined heat and power equipment, feed-in tariffs and grants for domestic renewable heat technologies that include wood biomass boilers and stoves (as well as solar panels and geothermal heat pumps).

Ireland's biomass contribution to energy, at about 2–3 per cent, is around a third of the European average and leaves Ireland ranked almost bottom among European countries. Reaching the 2020 potential would raise Ireland close to the leading position in Europe. It can be done by using natural indigenous resources, waiting to be put to work.

A GREATER GOAL

While Ireland's biomass resources provide the strong potential to help the nation move towards energy independence, they are not sufficient for full independence nor do they create a product with major export potential. The greater opportunity for Ireland in the biomass industry is the development of next-generation feedstock cultivation and conversion technologies that can be sold and applied throughout the global market. Ireland may not have the vast forests of Scandinavia, the steppe grasslands of Russia or the sugar cane plantations of the tropics but does have a wealth of biomass cultivation knowledge and technology in its biomass companies, agricultural cooperatives, farming support agencies and universities.

Several Irish companies have established a strong track record for providing bioenergy heating and power solutions to international clients. For instance, Clearpower, an Irish based bioenergy and organic waste management business, has developed heating, power, anaerobic digestion and waste recycling solutions for major international companies and government institutions and has even provided the biomass heating installation for the London 2012 Olympic Park. Dublin-based NTR plc has taken a key strategic position in global biofuels by investing in and helping to shape a leading bioethanol production group, NASDAQ-listed Green Plains Renewable Energy.

Decades of development and field testing at Ireland's agriculture and food development agency, Teagasc, in collaboration with farmers, has helped place Ireland on the world map for achieving some of the highest yields of wheat and grass on earth. The University of Limerick has developed unique technology in pyrolysis, fluid bed gasification and biochar application. The National University of Ireland, Galway has unparalleled technology in the field of enzymatic hydrolysis of cellulose and algae to produce fuel and high value co-products. University College Dublin and Trinity College Dublin have created mould-breaking technology in the fields of biodegradable plastic from biomass and yeast development. All these universities are actively working together to share their knowledge and build on each others' breakthroughs to create a world class centre of competence. They have attracted the interest and participation of the world's biggest energy companies. This intellectual property helps solve the 'food versus fuel' dilemma that reined in the early exuberance of biomass energy and can launch a new wave of sustainable energy growth. The knowledge is not limited by the availability and suitability of land in Ireland. It can make cheap, local and clean energy a reality wherever biomass grows in any corner of the world.

THE NUCLEAR OPTION

'A terrible beauty is born.'
– W. B. Yeats, *Easter, 1916.*

Some of the greatest mysteries of our existence have been solved by physicists in the last century. Ernest Rutherford and Niels Bohr shaped our knowledge of the building blocks of all life and matter – atoms – in the early 1900s. Albert Einstein determined the equivalence of matter with energy and revealed the enormous energy that could be unleashed from small particles. The splitting of an atom of uranium was first recorded in Germany by Otto Hahn and Fritz Strassmann in January 1939. A new horizon for energy and its manipulation came into view on this platform of knowledge.

World history and human fear aligned to drive the development of nuclear energy first towards military application rather than civilian benefit. In July 1939, four leading scientists, Szilard, Wigner, Sachs and Einstein, contacted US President Roosevelt and described the potential to create an atomic bomb that might stand up to the threat of German expansion. The US military began experiments under the codename 'Manhattan Project'. Scientists, academics and students throughout the US dropped their work and contributed to the initiative. The project was spurred on by the US entry into the Second World War from December 1941 and the belief that the Germans were rapidly developing their own atomic bombs. This latter stimulus later proved misplaced as little progress had been made in Germany. The work culminated with the first test of a bomb in New Mexico in July 1945 and the dropping of bombs on Hiroshima and Nagasaki, just twenty-one and twenty-four days later, respectively.

Scientists turned their endeavours to harnessing nuclear energy in a controlled way after the war. General Electric opened the first commercial electricity producing nuclear plant in 1960. A flurry of orders followed and a quarter of US electricity capacity came from nuclear energy within ten years. Worldwide adoption of peace-time nuclear energy initially grew quickly but became mired by controversies of waste, accidents, cost and public acceptance from the late 1970s. Today, 16 per cent of world electricity is supplied by nuclear plants in thirty-two countries. However, opinions on the risks and rewards of nuclear energy remain, like the fallout, heated and divided.

CORE CONCEPTS

Atoms are made up of a cluster of particles that can be described to behave like a tiny solar system. Particles in the core nucleus include positively charged protons and neutral neutrons that are bound tightly together. These are orbited by negatively charged electrons. The number of protons defines the element. For instance, hydrogen has 1 proton, carbon has 12 and uranium has 92. While atoms of an element always have the same number of protons, the same element may have different numbers of neutrons. These varieties of the same element are known as 'isotopes'. For example, the uranium-238 isotope has 92 protons and 146 neutrons, while the uranium-235 isotope has 92 protons and 143 neutrons.

The electrostatic repulsion of like-charged particles is so great that an even larger force of attraction is needed to keep an atom bound together in a compact structure. This attraction is a 'nuclear force' and it has an associated potential energy. If the atom is split or decays of its own accord, part of that nuclear energy is released. Some elements are naturally unstable and their atoms decay slowly over time. The best-known naturally unstable, radioactive elements are uranium and thorium. Several hundred new radioactive elements and isotopes can be produced, including plutonium, strontium-90, caesium-137 and iodine-131.

The decay of naturally unstable elements can be accelerated by the addition of man-made triggers in order to release the atomic nuclear energy. The energy released is proportional to the square of the speed of light, as outlined by Einstein ($E = mc^2$). Even a tiny mass may give rise to enormous energy as the speed of light is so great. It may be unleashed in the form of uncontrolled destruction or as controlled heat that can be applied to do useful work.

The natural or forced decay of atoms is accompanied by the radiation of atomic particles and rays. Alpha particles, which consist of two protons and two neutrons, are slow and heavy and may be stopped by a thin sheet of paper or the outer layer of skin. Beta particles, which are electrons, are lighter and quicker and lose their momentum in a few metres of air or a small thickness of glass or metal. Gamma rays are electromagnetic waves that may accompany the emission of alpha and beta particles and may be stopped only after many metres of air or several centimetres of a heavy metal such as lead.

All three radiation particles can displace the electrons in the atoms they strike, effectively destroying them. As a result they are extremely dangerous to life forms. Gamma rays are the most potentially destructive. A human can die from continuous exposure to intense radiation within hours as a result of the collapse of cell function throughout the body. Shorter exposure for a few minutes to intense radiation can result in prolonged death over months. Longer term exposure over the course of a year to small doses of radiation can result in cancerous tumours, blood disease and genetic defects passed on for multiple generations.

Decay of a quantity of radioactive material may take millions of years. The time taken for half a starting amount of unstable atoms to decay, the 'half-life', is 4.5 billion years for uranium-238, 700 million years for uranium-235, 10 million years for a lead isotope, 100,000 years for plutonium, and just a few hundred years for strontium and caesium isotopes. Radioactive decay usually follows a path of morphing from one unstable element or isotope to another until it finally rests in stability.

Natural radiation has infiltrated the environment from sources such as radon gas, which is formed from the decay of trace uranium and thorium in subsurface rock, and cosmic rays from deep space and solar flares. These sources

Radioactive isotope	Radiation type	Half-life
Uranium-238	alpha, gamma	4.5 billion years
Thorium-234	beta, gamma	24 days
Protactunium-234	beta, gamma	6.8 hours
Uranium-234	alpha, gamma	250,000 years
Thorium-230	alpha, gamma	80,000 years
Radium-226	alpha, gamma	1,600 years
Radon-222	alpha	3.8 days
4 short life isotopes	alpha, gamma	30 minutes
Lead-210	alpha, gamma	22 years
Bismuth-210	beta	5 days
Polomium-210	alpha, gamma	138 days
Lead-206	none	stable

Source: UN, Krauskopf

Radioactive decay path of uranium-238

of radiation have a relatively low intensity to which natural life is accustomed. Low-level radiation can also be applied benignly in medicine, from X-ray diagnostic tools to the treatment of cancer by destroying tumour cells with radiation. The spectre of intense bursts of life-altering radiation is raised by the introduction of man-made accelerated atomic decay for energy.

CAPTURING NUCLEAR ENERGY: HOW DOES IT WORK?

The common resource for nuclear energy generation is the uranium-235 isotope. When an atom of uranium-235 is struck by a slow-moving neutron, it becomes instantaneously excited by the additional energy. The atom distorts and assumes the shape of a dumb-bell before bursting apart into two roughly equal fragments. This *fission* is accompanied by the expulsion of atomic electrons that smash into adjoining atoms and molecules, generating heat. Additional neutrons are also released and these may strike and split other uranium-235 atoms, creating a chain reaction. If the chain reaction is uncontrolled, an explosion occurs, as at the strike location of a nuclear weapon. However, the chain reaction may be controlled for useful energy production in a nuclear reactor.

A controlled reaction is achieved by inserting an absorber of neutrons, such as boron, into the midst of uranium radioactivity in a reactor. At the core of the reactor are slender

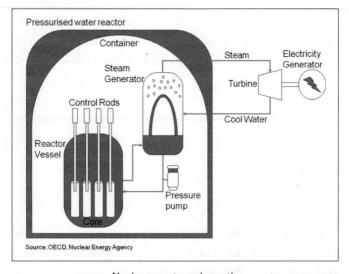

Nuclear reactor schematic

. fuel rods – metal tubes, several metres long and a centimetre in diameter – filled with uranium pellets and placed in a water tank. Absorber control rods may be adjusted to enable the reaction to proceed at a rate that matches energy requirements. The energy released in fission is used to heat water, creating steam, which turns a turbine for electricity generation.

Very little uranium-235 is needed to generate significant amounts of energy. Only 2.5 kg of uranium-235 is required to fuel a 2,000 MW thermal power plant each day. Several million kg of fossil fuel would be required for the same energy output.

WASTE

Nuclear power has a very low carbon impact. The amount of carbon dioxide emitted from nuclear power generation is at least fifteen times less than that emitted from fossil fuel power generation and even slightly less than biomass emissions. The comparison takes into account indirect carbon emissions from the mining of uranium and construction of nuclear plants. Furthermore, there are no emissions of sulphur, nitrous oxide or ash that accompany fossil fuel burning plants.

Less positively, the fission products of split uranium-235 include elements about half its atomic weight, such as strontium, caesium, cobalt and selenium in the form of radioactive isotopes. These continue to generate radiation and heat while they decay over several hundred years. A more serious form of high-level waste results from the decay of uranium-238, an isotope that is mixed in much greater bulk with the more fissionable uranium-235 when it is mined. It becomes unstable when it absorbs neutrons released in the fission process and commences its multi-billion year decay. Radioactivity in elements such as plutonium, neptunium, americium and curium is also triggered. Some of these elements remain intensely radioactive for tens of millions of years and present a major problem for safe waste management. Plutonium may be stripped from the rods and used as a nuclear fuel source itself but it is also a highly suitable material for bomb making and coveted by disparate groups throughout the world.

The high-level waste that accumulates in the fuel rods over time increasingly impacts the efficiency of energy production and the rods must be replaced every year or so. Some of the radioactive elements are stripped out in an acid bath. The spent rods and the hot acid solution must be managed for the length of their radioactive lives, which extend for about ten half-lives.

A 2,000 MW capacity nuclear power plant will create about 100 tonnes of radioactive waste and spent fuel each year. The amount of waste generated across the globe from nuclear energy production is about 18,000 tonnes each year. This would fill a sports stadium as big as Croke Park more than twice over.

The key challenge to radioactive waste management is not how to contain it safely but rather how to contain it safely for at least 250,000 years for most radioactive elements to decay,

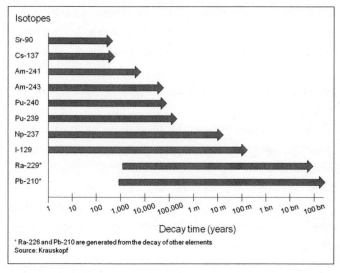

Lifetimes of radioactive waste from nuclear energy production

and for millions of years for a few residual elements. It is a question of maintaining control over time. Every government and interest group that presides through every future rise and fall of civilisation for ten million years or more must abide by the operating manual for radioactive waste management developed in our lifetimes in order to avoid future contamination. This time span far exceeds the duration of recorded human history – only 6,000 years. Is this responsibility and risk a reasonable burden to place on thousands of future generations?

The current proposal for dealing with waste is to dig a deep hole and bury it in a currently stable rock formation. This will allow for the 'out of sight' and 'difficult to reach' decay of the short-lived radioactive elements. The approach would be to sink a shaft to a depth of between 300 m and 1,500 m in stable formations such as crystalline rock, salt or compressed volcanic ash. Lateral tunnels would be excavated and metal containers of radioactive waste would be entombed. The tunnels and shaft would then be backfilled with gravel.

However, confidence in this proposal is so uncertain that no geological repository has yet been brought into service, despite almost fifty years of nuclear power and waste generation. The approach accepts with certainty that metal waste containers will eventually corrode and shifts in geology will eventually release the long-lived radioactive elements. The earth's crust is like a flimsy veil on a core of molten fluid and it is constantly shifting, shearing and fracturing. The approach is based on the hope that released elements will have decayed enough and only acceptable amounts of radioactive atoms will reach groundwater, end up in the food chain and ultimately reside inside the bodies of humans and animals. It is a hope based on geological reconstructions of the past but not the future.

SAFETY

Safe production of energy is the primary concern of any nuclear plant operator. Risks to safety are minimised by several mutually reinforcing factors. The plant is designed to incorporate accident prevention systems. Plant equipment is fabricated and inspected to internationally agreed standards that seek to eliminate any defects that might lead to flawed performance. Physical barriers are constructed with a view to containing any releases in the event of an accident. In operating mode, the slow emission of neutrons and the fact that rising temperatures expand the size of neutron absorbers inherently help to maintain control of the reaction rate. Control rods that absorb neutrons can be lowered into the reactor core to stop the reaction in an emergency and these are set to 'fail safe' or automatically lower if normal operating conditions are breached. In the event of a core failure, high-pressure injection systems are triggered to flood the core with coolant and coolant units are started throughout the plant. If a release to the environment occurs, a planned chain of public alerts and contamination management plans are set in motion.

Despite best intentions, several serious accidents have occurred at nuclear power plants including Windscale (1957), Three Mile Island (1979), Chernobyl (1986) and Tokaimura (1999).

Windscale

Britain was determined to join the US as a front runner in the global nuclear race after the Second World War. The Windscale plant in Cumbria (now the location of Sellafield) was built in the late 1940s with the purpose of producing plutonium from uranium to build Britain's first atomic bomb. Plutonium production at the plant was initially slower than required by the government's nuclear programme. It was decided to increase operating temperatures above design specifications and reduce the size of aluminium safety cartridges that contained uranium in order to speed up production.

The nuclear race intensified after the US exploded the world's first hydrogen bomb in 1952. Winston Churchill publicly committed Britain to developing the 'H-bomb' and put pressure on the managers at Windscale to further accelerate production. The plant began to creak under the strain. The material properties of graphite that surrounded the reactor started to transform under stress. Radioactivity levels in the local vicinity began to rise due to small leaks. As an international nuclear test ban deadline in 1958 loomed, production at Windscale was ordered to be increased by 500 per cent. Safety procedures were curtailed to accommodate the political demands. The size of the aluminium safety cartridges was further reduced, increasing the risk of fire.

On 7 October 1957, operators increased the temperature of the reactor core to enable a release of stored energy. The release of energy did not work as expected. Operators decided to increase the temperatures a second time and unknowingly started a small fire in the core. After three days, operators noted that the core temperature was rising steadily. They decided to blow air into the core to cool it. This fanned the flames and the fire spread rapidly throughout the core and out of control. The pumped air also drove radioactive dust and smoke out of the plant chimney and across the surrounding countryside. No public warning was issued to the local village of Seascale. Senior managers at the plant phoned the local school and asked that their children be sent home. Several rang their spouses and told them to drive away immediately. Half of the village residents, those who had been informed by family members working at the plant, left.

The most senior manager on duty subsequently reported in an interview that there was no safety plan for dealing with fires and that they had to 'play it by ear'. Workers went to the reactor core and tried to push uranium cartridges free of the flames using nearby scaffolding poles. It is reported that local police were dispatched to the village cinema, where they stopped the film and 'enlisted' men to go to the plant and assist in the manual removal of uranium cartridges.

Water hoses were turned on the fire, introducing a significant risk of causing a steam or hydrogen explosion that would likely have killed workers in the plant and caused regional nuclear fallout for hundreds of kilometres. No explosion occurred though the water also failed to quench the flames. Finally, operators decided to turn off the air fans and the fire was starved of oxygen and began to die.

The public was later informed by BBC broadcast that an overheating incident was under control and reassured that wind was now carrying radioactive dust away from land and over the Irish Sea. Concerns were raised of the increased risk of thyroid cancer for babies, and milk produced in a radius of 200 km was condemned and poured into the Irish Sea. There were no immediate deaths, though it is estimated that some 240 cases of cancer were caused.

Three Mile Island

At 4 a.m. on 28 March 1979, the Three Mile Island plant in Pennsylvania was operating normally when a valve to the steam generation system was left closed for eight minutes by mistake. A subsequent build-up of steam pressure caused a relief valve to burst open and cool water that fed the steam generator was released to an adjacent tank. This cool water contained radioactive contamination. The tank filled quickly and ruptured under pressure, releasing its contents into the building and the contaminated water found its way to drains. Operators decided to close the cool water pipes to prevent further release. As a result, the core heated up in the absence of sufficient coolant and an uncontrolled chain reaction occurred. The top 1.5 m of the core melted and collapsed, damaging the reactor, and radioactive gases were released. The containment tanks overflowed and some radioactive gas was released to the atmosphere. While most leakage was contained within the plant, the fraction that did escape was detected at a distance of some 100 km. At this outer area, radiation levels were similar to that of an X-ray machine. Closer to the reactor, many people were evacuated and special care was given to pregnant women in the region. There were no fatalities and increased cancer levels in the region were estimated to be low.

Chernobyl

During 25 April 1986 and into the first hours of 26 April, an emergency simulation test was conducted at the Chernobyl nuclear plant near Kiev in the then Soviet Ukraine. The objective was to determine how electricity and coolant supply to a reactor would be affected if nuclear power levels suddenly fell during a shutdown. The first step of the test was to reduce the nuclear power to 30 per cent of its normal operating level. The power was mistakenly lowered to about 1 per cent by the test operators. Against all the rules, the operators pulled most of the control rods out of the core in an effort to raise the power quickly again. The sudden variations in operation led to an overheating of the coolant and the creation of pockets of steam in the coolant pipelines. With the control rods removed and the coolant system malfunctioning, the nuclear power flashed to ten times its normal operating level. The enormous surge of energy pulverised the core fuel. Temperature, pressure and steam built up and an explosion blew a hole in the roof of the building. The influx of air led to additional chemical eruptions that sent molten material hurtling into the environment and starting some thirty fires outside the building. The searing heat vaporised the core fuel and released an enormous radioactive cloud that swept across Eastern Europe towards Scandinavia and nuclear rain was later detected as far west as Ireland. The first international notification of the fallout came when increased radiation levels set off alarms at a Swedish nuclear power plant on 28 April. Higher radiation levels were subsequently detected throughout the globe.

Some 230 operating employees and emergency workers were brought to hospital with acute radiation sickness, of whom thirty-one died within three months and a further fifteen died over several years. The first firefighters on the scene picked up pieces of the scattered core and died from radiation sickness within weeks. One firefighter said that the radiation tasted like metal and described a sensation of pins and needles all over his face. Thyroid cancer killed a further nine local children. Over 330,000 people from towns and villages in the vicinity were exposed to intense radiation. They were informed more than twenty-four hours after the explosion. Over 135,000 people were permanently evacuated from a 30 km exclusion zone. Millions more people beyond the immediate zone were exposed to radiation. Radiation levels in Europe outside the Soviet Union increased to several times normal background levels. Some 100,000 square km of land was contaminated and a ban on food imported from the Soviet Union was subsequently enforced by several European countries.

Large amounts of sand, rock, boron and lead were dumped from helicopters into the reactor, which continued to burn for two weeks. One helicopter struck a nearby crane and crashed, killing its two-man crew. Some time later, the earth beneath the plant was excavated and replaced with concrete to try to prevent leakage to groundwater. The buried core is expected to continue to generate heat until 2075 and radioactive decay at the site will continue for several million years.

Tokaimura

Japan experienced three serious nuclear accidents in the late 1990s. In 1995 a coolant leakage and fire took place at a nuclear reactor in Monju, 330 km west of Tokyo. In 1997, thirty-five people were exposed to radiation from a fire and explosion in a waste processing plant in the village of Tokaimura, 130 km northeast of Tokyo. The most serious accident occurred in a nuclear fuel factory in Tokaimura in 1999.

On 30 September 1999, operators mixed nuclear fuel components in a stainless steel tank rather than the required mixing equipment in order to save time and mix larger amounts. The operators were following an internal company manual that had been drafted illegally and violated basic safety rules. The uranium fuel began a self-sustaining chain reaction, which emitted intense gamma and neutron radiation. The chain reaction continued for twenty hours before being brought under control. The plant ventilating fans blew radioactive air into the local village throughout that period. Five hours after the start of the accident, local residents were evacuated. Twelve hours after the accident some 300,000 people within a 10 km radius were asked to stay indoors for a day. One hundred and nineteen people were identified as having been exposed to radioactive contamination. Two plant workers, who had been wearing T-shirts at the time of the accident, died within several months.

Risk remains high

Major lessons have been learned from these incidents and others – in particular the need to minimise the likelihood and impact of human error. Improved design, planning and operation have led to a large reduction in the probability of accidents at nuclear energy plants. Some proponents of nuclear power point to the fact that the actual loss of life and accident rate in the nuclear industry is low compared to other industries and activities such as mining or transport. They suggest that media tend to exaggerate the probability of accidents and that public opinion is influenced by unjustified fear. However, 'risk' is simply a product of both probability and consequences. Even if the probability of an accident is indeed low, since the consequences are deadly serious, the risk always remains high.

COSTS

A key challenge to nuclear power is the dual cost of construction and decommissioning. Plants can take up to ten years to build and budget overruns of up to 200 per cent have been experienced in many recent construction projects in the US, India and Europe. Costs of decommissioning are unpredictable owing to the nature of waste handling. The cost for decommissioning the first plants in the UK is estimated to exceed €100 million, much of which will be carried by the taxpayer. Site decommissioning is estimated to take up to fifty years, roughly the same time as the useful life of a producing plant. The operating costs are broadly equal to the costs to operate a fossil fuel plant when carbon costs are taken into account but slightly more expensive than onshore wind energy generation costs.

FUSION ASPIRATIONS

Fusion is a more benign cousin of fission. It could use an effectively unlimited source of fuel from ocean water and create very limited pollution if the technology were made feasible. Fusion occurs when two light elements join together to create a heavier element and liberate energy in the process. Unfortunately, the repulsion of like-charged protons is so great that enormous energy is required to bring atoms close enough for fusion to occur. Nuclear fusion is the primary source of energy in the sun and stars. Artificial fusion on earth has yet to achieve an energy output greater than the input required.

There are two possible approaches to achieving fusion. In the first, hydrogen atoms are accelerated towards each other or bombarded by a laser with such massive energy that the repulsion forces are overcome and fusion occurs. An alternative approach is to heat a plasma of hydrogen to temperatures as high as 150 million °C, several times that of the core of the sun. The plasma is kept in place by a superconductor magnetic field long enough for fusion to occur. However, the energy output from resultant fusion in either approach is estimated to be just two thirds of the input required and most materials and equipment would melt and vaporise at the high temperatures.

One promising fusion project, called 'Iter' (the Latin for 'way'), has been established in southern France with joint collaboration by governments and scientists of the EU, China, India, Japan, South Korea, Russia and the US. The prototype technology uses magnetic coils to confine a gaseous mixture of deuterium and tritium, two varieties of hydrogen, which is brought to a fusion temperature of 200 million °C. The Iter project costs rise by the billion and its completion date continues to be pushed back year after year. A further potential development is the creation of a hybrid fission-fusion reactor, where the energy from fission is directed to create the conditions required for fusion. Such a hybrid reactor may be capable of converting waste products so that they decay over hundreds of years rather than tens of thousands of years, thereby reducing one of the key shortcomings of conventional fission.

The search for the holy grail of energy-positive fusion continues, driven by the allure of abundant source material in the form of deuterium from the oceans, and the elimination of harmful waste. Continued research and testing builds on decades of work to achieve the aspiration of clean energy from fusion but it will be many more decades before the aspiration comes close to reality.

HOW MUCH NUCLEAR ENERGY IS AVAILABLE?

The amount of available nuclear energy is dependent on the finite reserves of suitable uranium. Although uranium is ubiquitously distributed in the earth's crust and even in tiny amounts in sea water, it may only be mined economically where it is found in high concentrations. There are a handful of locations where the resource is recoverable. The total amount of known recoverable uranium is about 5.5 million tonnes. Some 23 per cent of this is found in Australia, while a further 25 percent is found in Kazakhstan and Russia. South Africa and Canada each own 8 per cent and smaller amounts are scattered in the US, Brazil, West Africa and Asia.

The current annual usage of uranium is 65,000 tonnes. Consumption is rising at 2 per cent each year as new plants are brought online. At these rates, current known resources will last only for a further fifty years. If nuclear electricity supply were to double, known resources would run out in half the time. Nearly €600 million is invested each year in exploration for new reserves. As uranium prices rise, more will be spent on extracting it from challenging deposits. Industry advocates suggest that 'undiscovered' reserves might sustain current nuclear electricity supply for over 200 years. The availability of nuclear fuel may also be extended by refining spent uranium into fissionable plutonium, reprocessing nuclear weapon stockpiles, developing 'fast neutron reactors' that use less uranium or creating new technology to harness other unstable elements, such as thorium. In any event, the resource is finite and will experience the same peaking dynamic as fossil fuels in due course.

There are 443 nuclear power plants around the world providing 16 per cent of the world's electricity needs, including 21 per cent of electricity needs in OECD countries. The countries that use most nuclear power are the US, France, Japan, Russia and the UK. In France, 78 per cent of electricity generation is nuclear. Several countries, including The Netherlands, Belgium, Austria, Spain and Italy, have voted to restrict or phase out nuclear energy on the basis of accident risk and radioactive waste hazards. Nuclear energy is vetoed or illegal in many countries including Australia, Austria, Denmark, Norway and Ireland.

NUCLEAR ENERGY IN IRELAND

The global trend towards nuclear power in the late 1960s influenced several Irish energy policy makers. In 1968, four nuclear power plants were proposed to be built at Carnsore Point, County Wexford. A groundswell of public and environmental group opposition hampered plans throughout the 1970s. Some 10,000 people descended on the Point to join protests and enjoy free concerts in the summers of 1978 and 1979. Christy Moore, Clannad and Chris de Burgh headlined the music line-up. Sandwiches cost 15 pence and a cup of tea was 10 pence. Environmentalists, academics and politicians took their turns at a microphone decrying the plans for nuclear energy to rain-soaked listeners. It was an Irish Woodstock and the government of the day also listened from a distance to the peaceful protests that were a touchstone of national sentiment. Nuclear plans were abandoned eventually in 1981 and fourteen wind turbines now stand on the site.

Today, there is a statutory prohibition on nuclear power generation in Ireland. The debate continues and state agencies, semi-state bodies, trade unions and lobby groups, including the Electricity Supply Board, Forfás, the Irish Business and Employers Confederation, the Irish Congress of Trade Unions as well as politicians have turned up the volume in the light of energy and climate concerns. There are several key arguments on either side of the nuclear debate in Ireland. Public opinion will most likely remain the ultimate determinant on whether one or more nuclear plants will ever be built.

Pros:

The primary arguments in favour of nuclear energy in Ireland include the low carbon impact of an operating plant and a reduction in national dependency on fossil fuels.

Carbon dioxide emissions could be reduced by roughly 5 million tonnes per year if a single 1,000 million W nuclear power plant, providing a fifth of peak electricity demand, were to substitute the worst polluting plants including coal-fired Moneypoint. This would result in carbon permit savings of some €100 million every year, at an estimate of €20 per tonne of carbon dioxide. A saving in the overall social cost of carbon as defined by the International Panel on Climate Change would amount some €300 million every year. Nitrous oxide, sulphur dioxide and toxic heavy metal emissions would also be removed. Nuclear power would enable Ireland to improve its international ranking in the league of greenhouse gas emissions and help meet Kyoto obligations.

If the threat of climate change is accepted, then a real option for averting economic, social and environmental damage cannot be ignored by any country. The ability to replace polluting fossil fuels with renewable energy sources is challenged by issues of cost competitiveness and the slow pace of substitution – it has taken twenty-five years for wind energy to substitute some 1.5 per cent of fuel supply. What if practical introduction of renewable sources continues to fall short of aspirations? What if oil and gas prices soar further in response to peak dynamics and the energy industry turned to coal, a more damaging contributor to climate change, as a cheap substitute? Is it worse to leave a legacy of radioactive waste from nuclear power or a legacy of catastrophic climate change due to fossil fuel emissions and global warming? Should nuclear languish on a blacklist in all circumstances?

Nuclear energy would provide a diversification from Ireland's heavy dependence on fossil fuels. Although overall fuel import dependency would not improve, diversification helps to spread risks to pricing and security of supply. Additional security may be achieved as uranium can be stockpiled more easily than other fuels.

Cons:

The primary arguments against nuclear energy in Ireland include the issues of waste management, safety, inability of the grid to accommodate even the smallest available size of nuclear plant, dependency on a finite fuel source, cost and considerable public opposition.

Ireland has no radioactive waste facility nor suitably identified subsurface geological formation for long-term storage. Therefore, Ireland would be doubly reliant on foreign support – dependent on secure and continued import of uranium fuel, substituting one finite imported fuel for another on the one hand, and secure and continued export of waste for disposal, possibly to France, on the other.

A nuclear plant would introduce significant safety issues in the form of the risk of radioactive material leakage and the risk of a terrorist attack. A direct hit by a passenger jet could create conditions similar to the Chernobyl fallout, mass national evacuation and a significant increase in the number of cases of cancer.

The Irish electricity grid cannot readily accommodate plants that are as large as the smallest available and economic-sized nuclear plant, some 1,000 MW, due to the small size of our national demand, which is some 5,000 MW at peak. Nuclear power cannot be switched on and off at short notice in response to the fall and rise of demand because the ramp-up of a nuclear plant takes several days. Therefore, a nuclear power plant would need to be in continuous operation, as a 'baseload' source of supply. However, nuclear plants do 'trip' normal operating conditions on occasion and shut down. While a large electricity grid could absorb such supply shocks, a small grid such as Ireland's cannot. A replacement supply that could be ramped up almost instantaneously, such as gas-fired generators, would be required to fill the sizeable and sudden void. This would necessitate a 'shadow' generating capacity at significant cost. Although an interconnector to Britain could provide emergency supply in such a circumstance in theory, the commercial and operating conditions of the 500 MW interconnector between Wales and Ireland do not allow for this.

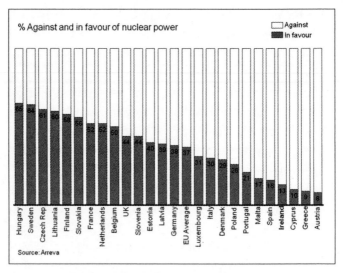

Public opinion in Europe on nuclear power

The capital cost of building a 1,000 MW nuclear power plant would be in the range of €4–€5 billion, based on international benchmarks. A similar energy capacity provided by natural gas or wind would cost in the region of €900 million or €1.2 billion respectively.

Public opinion is strongly opposed to the establishment of nuclear power in Ireland. Opinion drivers include the desire to maintain a 'nuclear-free' national image and antipathy fostered over several decades to the Sellafield (formerly Windscale) reprocessing plant in Wales.

Despite clear public opposition and a statutory prohibition, nuclear energy will fuel Irish homes and businesses through the back door. The 500 MW electricity interconnector between Ireland and the UK will enable the export of wind-generated electricity, lower consumer prices and a smoother overall supply profile. It will also facilitate the importation of nuclear-generated electricity. On occasions when electricity is flowing from the UK at full demand, some 2 per cent of Ireland's total electricity will be nuclear. This will rise to some 3 to 4 per cent when new UK nuclear plants come on stream. Ireland is no longer a nuclear-free zone.

ENERGY AND CARBON EFFICIENCY

'. . . the South Dublin Guardians, notwithstanding their ration of 15 gallons per day per pauper, supplied through a 6 inch meter, had been convicted of a wastage of 20,000 gallons per night . . . thereby acting to the detriment of another section of the public, selfsupporting taxpayers, solvent, sound.'
– James Joyce, *Ulysses*.

There is another way to wean ourselves from the addiction to carbon fossil fuels that complements switching to alternatives. It is applying whatever energy resource we use in an efficient way. We can see, touch and smell a sod of peat, a flow of petrol or a cresting wave. Energy efficiency is less tangible but its effects are just as impactful. In fact, using energy efficiently is the simplest, fastest way to reduce greenhouse gas emissions and energy dependency and can also achieve significant savings.

Leading OECD economies have managed to reduce energy demand by about 15 per cent on average over fifteen years as a result of introducing energy efficiency policies. However, it is estimated that about two thirds of the value that could be captured by efficiency measures has yet to be tapped. The EU has estimated that a fifth of all energy demand could be removed using efficiency activities, yielding savings greater than the costs required. The accompanying reduction in carbon dioxide emissions would be twice the full reduction required under the Kyoto protocol.

There are two approaches to achieving energy efficiency: 'use less' and 'do more with what we use'. Both approaches require a change in mindset that acknowledges that energy is a real and precious asset rather than something to be taken for granted and always available at the flick of a switch. This sounds simple, although it is easier said than done. We generally regard doing any legal thing as an unquestionable right as long as we can afford it – if we can pay for it then we have earned the right – without giving attention to the energy required. We have all become used to accepting waste in systems and behaviours. We live with inherent inefficiency in almost everything we do.

For example, incandescent light bulbs convert just 5 per cent of energy into useful light. The balance of 95 per cent of energy used is lost as dissipated heat. That waste is rarely considered when we turn on a light. Cars are also a highly energy inefficient means of transport, though we take their use for granted. The current fuel efficiencies of petrol and diesel engines are only 30 per cent and 45 per cent respectively at best. The remaining energy is wasted in the process of converting the chemical energy of the fuel to

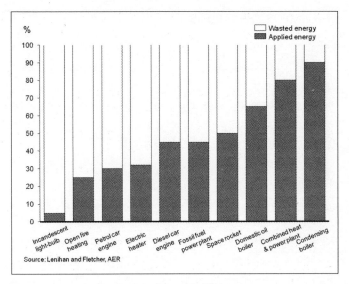

Some typical energy efficiencies

kinetic energy. It gets worse: just a third of this converted kinetic energy is actually put to work in moving the car when tyre friction, vibrations and other losses are taken into account. It gets worse again: fuel consumption is proportional to weight at a constant speed. Therefore, 95 per cent of the energy used is applied to transport 1,500 kg of metal and material and only 5 per cent is applied to transport a 75 kg passenger. This does not even take the energy required to produce the car into account. Despite the phenomenal inefficiency and waste, most people who own a car cannot fathom life or work without the personal freedom, convenience and ability to commute that it affords.

SELF-IMPOSED BARRIERS

The natural familiarity with accustomed ways of doing things and unthinking acceptance of inherent inefficiency makes it hard to change to more efficient practices. We battle the inertia of reluctance, resistance and lethargy. It often feels like too much hard work to change – why bother if we are already getting what we want and need?

The same constraint to change is a feature of systems and infrastructure in place and sometimes literally set in concrete. Who is going to replace the existing vast network of oil pipelines, terminals and fossil-fuel filling stations and pay for smart electric networks with battery charge points or hydrogen highways and dispensers (notwithstanding a small handful of installed 'proof of concept' sites)? Which car manufacturer will lead the charge in mass-producing hydrogen cars in a yet unproven market without supporting infrastructure?

The availability of capital is a constant barrier to improving energy efficiency. Despite the prospect of a positive payback, banks and equity providers simply do not always sign up to the promise of future earnings in exchange for significant cash today. Competing uses for funds, uncertainty about future returns and conventional financial modelling and discount rates that do not take efficiency savings into account all conspire to shackle opportunities from taking effect.

The lack of an empowered or enlightened decision maker who manages energy use results in continued energy waste. Who is in charge of turning off the air conditioning, lights or heating at night in an office block? Who cares if several thousand lights are left on in a closed university library overnight? Why bother to upgrade a heating system to increase its efficiency and lower fuel consumption when the one in place gives all the heat needed and deciding how to improve things is not part of one's job specification but another person's job at headquarters?

REWARDS FROM EFFICIENCY MEASURES

Overcoming the barriers and embracing energy efficiency measures can achieve enormous rewards. Energy consumption and bills can be lowered by a third in most businesses and homes. Greenhouse gas emissions and fuel dependency can be reduced in tandem. Some of the top energy efficiency measures are described below. All activities provide a 'positive payback' for any investment made. Savings of at least 10 per cent can be achieved at no cost at all. Any reduction in electricity consumption yields a double dividend as losses of some 50 per cent occur in electricity generation.

Lighting: Between 20 and 40 per cent of electricity costs for buildings arise from lighting. More than a third of these costs can be avoided.

▶ Switch to energy-efficient light bulbs. Change incan descent bulbs to long-life halogen lamps, compact fluorescent lamps (CFLs) or light-emitting diodes (LEDs). Changing a 100 W incandescent bulb that is lit for four hours a day to a CFL can save almost €20 per year. Changing fluorescent tube bulbs from 'T12' to more efficient 'T8' will save 10 per cent of costs while providing the same light. More than 700 coal-fired power plants, over a quarter of the coal-fired plants in the world, could be closed if all light bulbs were upgraded for efficiency. The share of global electricity demand for lighting would fall from 20 per cent to 7 per cent.

▶ Install presence detectors that turn lights off in rooms which might be periodically empty.

▶ Add light meters and dimmer switches that enable lights to be dimmed when there is sufficient natural light in a room.

▶ Design new buildings to take greater advantage of daylight and light-saving equipment

Heating: Up to 60 per cent of total energy costs can be attributed to heating of air and water. A third of these costs may be clipped from energy bills.

▶ Check that a boiler thermostat is at the right temperature to ensure unnecessary heat is not being generated. Dropping the thermostat setting by 1 °C can cut a heating bill by 10 per cent.

▶ Add a heating control system that measures ambient and room temperatures and aligns heating with actual conditions, avoiding over-heating and switching on and off when required.

▶ Install a condensing boiler to increase efficiency to 90 per cent. Old and inefficient boilers can waste up to half of the fuel burned for heat. Condensing boilers work by recycling heat from flue gases that are created as the fuel is burnt and minimising the heat escaping through the vent.

▶ Fit thermostatic valves to individual radiators. These enable a user to set a desired temperature rather than overheating a space.

▶ Insulate hot water tanks and pipes to stop expensive waste. The cost of a lagging jacket on a hot water tank will be paid back from energy bill savings within three months.

▶ Recover heat from industrial processes such as compressors, engines, kilns, ovens and refrigerators and reduce the need for heat from a central boiler.

▶ Use solar energy and air or ground-source heat pumps to increase base temperatures and heat water.

Air conditioning: Air conditioning is energy hungry and can double the energy consumption and emissions of a building.

▶ Open the windows to cool a house or building when the temperatures allow. Increase night ventilation of natural air instead of using air conditioning to help keep a building cool during hot periods.

▶ Ensure that controls of timing and temperature are set to 'automatic' and match the conditions. Some 90 per cent of air conditioning systems are estimated to be at the incorrect settings or run all night and during holidays.

Building fabrics: Over half of heat is lost through the building material (with the balance lost through draughts or ventilation). Choosing the right materials or adding support to an existing building can reduce energy consumption and bills significantly.

▶ Insulate. Up to a third of all heat is lost through walls. Wall cavity insulation can reduce this by 66 per cent. A quarter of heat is lost through an un-insulated roof. Proper roof insulation of at least 200 mm thickness will reduce this by 90 per cent. Good insulation can reduce the overall heating bill by a phenomenal 45 per cent.

▶ Replace old windows with double glazing to reduce the heat loss when it is colder outside and add blinds or shades to minimise heat ingress when in it hotter outside. Close the curtains at night and ensure curtains are not blocking radiator heat from entering the room.

▶ Add draughtproofing to doors, windows and walls to ensure rooms are airtight.

Monitoring: Measuring and monitoring energy use enables a better match of consumption with actual needs.

▶ Install a smart meter, which monitors and analyses energy consumption. Homes and businesses that have installed a smart meter tend to reduce energy consumption by an average of 3 per cent immediately as a result of consuming energy with raised awareness. More focused analysis can reduce energy consumption further. For example, energy consumption can be compared across multiple similar office or production locations to ensure that all are consuming efficiently; data can indicate if equipment is working efficiently; unusual patterns of consumption over time can be identified; and billing accuracy can be checked. Smart meters can also include a net-metering feature enabling a user to sell micro-generated electricity – for example, from a small wind turbine, solar thermal system, geothermal system or photovoltaic panel – back to the grid.

Office equipment and information and communications technology (ICT): Electricity costs of office equipment can be reduced by two thirds through more efficient management. ICT uses up to 10 per cent of electricity in modern societies and its consumption rate is growing every year. The number of devices joining networks doubles every 2.5 years and the amount of data created doubles every 1.5 years.

▶ Add timers to computers, printers and photocopiers to power down at night.
▶ Unplug chargers and equipment that draw small currents.
▶ Turn off machines that are in stand-by mode when they are not needed and save 20 per cent of the energy consumption.

▶ Take advantage of cloud computing and outsource hardware and software applications to a third party to optimise the use of servers and reduce local ownership costs.

Household appliances: Using appliances efficiently is the single most effective way to reduce energy use and bills in the home.

▶ Turn off appliances in stand-by mode. A television left constantly in stand-by mode will use about half as much energy compared to when it is turned on.

▶ Hang clothes to dry when possible to avoid using a tumble dryer.

▶ Boil only the amount of water that is required in a kettle. Why fill it for one cuppa?

▶ Keep the oven door closed while cooking – every time the door is opened to 'check on progress' a fifth of the heat is lost.

▶ Use a microwave to reheat rather than an oven.

▶ Put the lid on pots when boiling to trap the heat and boil more efficiently and quickly.

▶ Clean frost from fridges whenever it builds up. Let cooked food cool before putting it in a fridge.

▶ Turn lights off in rooms that are not in use.

Motoring: More efficient driving saves fuel and money, reduces wear and tear on the car, extends a car's life and is safer.

▶ Slow down and keep to the speed limits. Driving at 120 km per hour uses 25 per cent more fuel on average than driving at 100 km per hour.

▶ Check that tyre pressure is correct. Pressure that is 20 per cent lower than the recommended level increases fuel consumption by 3 per cent, reduces the tyre life by a third and increases the risk of sudden failure and a fatal accident.

▶ Drive smoothly as rapid acceleration and braking waste fuel.

▶ Shift gear up before revolutions reach 2,000 in a diesel car or 2,500 in a petrol car.

▶ Ensure a vehicle is serviced regularly.

▶ Carpool to work or for school runs when possible.

▶ Use public transport when possible.

▶ Walk or cycle instead of driving.

ENERGY AND CARBON EFFICIENCY IN IRELAND

When all energy efficiency activities that could be undertaken in homes, businesses, public sector and transport in Ireland are assessed for economic viability and practicality, some 33,600 GWh of energy could be saved. This is equivalent to almost a fifth of the total energy used nationally. The savings include more than one fifth of peak electricity demand, the most expensive electricity. Furthermore, up to 8 million tonnes of carbon dioxide emissions could be avoided, saving €160 million in annual

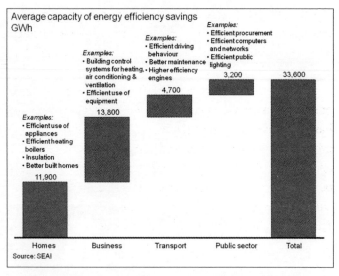

Potential for energy efficiency savings in Ireland by 2020

carbon permits, at a cost of €20 per tonne, and €480 million per year in the overall social cost of carbon, as defined by the IPCC.

Perhaps the most extraordinary boon of energy efficiency is the major financial payback that can be achieved. Over 1 million of Ireland's 1.7 million homes stand to benefit from investing in energy efficiency measures and could cut up to €700 from energy bills each year. On a national level, an average investment of €65 million each year to 2020 would achieve a 20 per cent energy efficiency improvement and would yield an average payback to the economy of some €340 million each year over the same period. This is a huge return on investment. If this were solely a private sector opportunity, investors would work day and night to exploit it. The fact that it is a distributed opportunity – one that belongs in divided rations to every householder, business and service provider in the country – somehow makes its exploitation less efficient.

A comparison of activities that can reduce carbon dioxide emissions in Ireland shows that energy efficiency activities have the potential to save significant money compared to all other activities. The graph below shows nineteen different approaches to reducing carbon dioxide emissions that have been analysed by the Sustainable Energy Authority of Ireland and McKinsey and Company management consultants. Those activities to the left of the y-axis actually save more

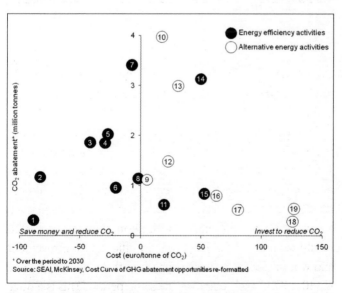

Comparison of opportunities with the potential to reduce carbon dioxide emissions in Ireland (40 per cent of opportunities save more than they cost)

money than they cost to implement over the period to 2030. It can be seen that almost all energy-efficient activities save money. For example, improving the efficiency of electrical appliances (1) saves the most money, while better built buildings (7) saves money and has almost the highest impact on reducing carbon dioxide.

The activities analysed by the Sustainable Energy Authority of Ireland and McKinsey were:

❶ Efficient electrical appliances (e.g., higher efficiency refrigerators, cooking, household and office appliances; more efficient computer and server management systems; low-energy consumption equipment; automatic shutdown controls).

❷ Efficient light bulbs and controls (e.g., switching incandescent bulbs to CFLs and LEDs in homes and offices and adding motion detectors and dimmers; ensuring the 400,000 street lights and 15,000 traffic lights in Ireland have energy-efficient bulbs).

❸ Improved efficiency petrol cars (e.g., ensuring cars achieve a limit of 130 g carbon dioxide emissions per km; conforming to the EU target of 95 g carbon dioxide per km by 2020).

❹ Basic insulation (e.g., attic and wall cavity insulation; draughtproofing windows and doors).

❺ Retrofitted heating and air conditioning and more efficient systems and maintenance (e.g., condensing boilers for heating; replacing electric resistance

heaters; better heating controls; improved maintenance; servicing the 30,000 AC systems with output
over 12 kW in 14,000 locations in Ireland).

6 Efficient cement production (e.g., clinker substitution
by limestone, blast furnace slag or pulverized fly ash;
using alternative fuels such as waste and biomass).

7 Better built buildings (e.g., improved design and
building orientation; efficient materials for construc -
tion of walls, roof, floor and windows; installation of
high-efficiency heating and ventilation systems;
reduced floor space; addition of energy controls).

8 Efficient farm management (e.g., grassland and crop-
land nutrient/slurry management; increased planting
of biomass; organic soil restoration; livestock grazing
period management).

9 Liquid biofuels (e.g., first and second generation
bioethanol and biodiesel).

10 Onshore wind generated electricity (including a
gradual switch from peat to wind).

11 More efficient industry processes (e.g., process
optimisation in manufacturing, power, cement).

12 Hybrid cars.

13 Forestation of pastures.

14 Replacement of Moneypoint coal plant with carbon
capture and storage plant.

15 Advanced insulation (e.g., high-efficiency windows
and doors and the use of materials to facilitate energy
to heat indoor spaces exposed to sunlight.

- ⑯ Offshore wind-generated electricity.
- ⑰ Biomass power (including co-firing biomass with coal).
- ⑱ Ocean energy (e.g., wave and tidal).
- ⑲ Electric vehicles (e.g., aggressive fleet penetration of 10 per cent by 2020 and 20 per cent by 2030).

CHOICES

It is helpful to know how personal choices such as means of transport and even diet affect the intensity of energy use and greenhouse gas emissions. Public transport generally results in lower carbon dioxide emissions and greater energy efficiency, especially compared to cars with single occupancy. It is interesting to note that even cycling and jogging emit small amounts of carbon dioxide as the human body combusts food for energy.

Animals raised for human consumption contribute over a tenth of global greenhouse gas emissions and require intense use of resources and energy, from farm machinery to feed production to transport and packaging. Cattle are the biggest contributors. A cow will eat up to 100 kg of feed and over 100,000 litres of water will be used for every kilogram of beef produced under intense farming conditions. Greenhouse gas emissions created from the production of every kilogram of beef are equivalent to driving a car for over 250 km. Over the

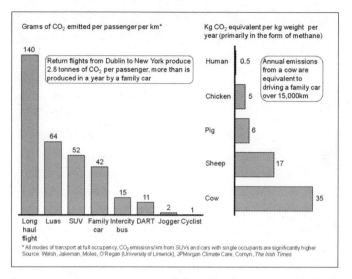

Comparison of greenhouse gas emissions from transport and animals

period of a year, a cow's belching and flatulence will contribute more to the greenhouse gas effect than a car, taking into account the fact that methane is twenty-three times more harmful than carbon dioxide. In total, up to fifty times more energy is used in producing beef than ends up as protein energy in steak on a plate.

Making personal choices to use energy efficiently results in a less expensive and healthier lifestyle. Choosing to devise and implement energy efficiency activities on a national level, as the government is committed to doing, creates major benefits for the economy and all citizens. However, changes to national energy efficiency policy have been incremental and low impact to date. A stronger government approach and bolder changes should be adopted in order to enable Ireland to reap the benefits. For example, building standards for energy efficiency in Ireland significantly lag behind those in most other European Union countries. Why should Ireland not take a position of leadership and adopt the highest building energy efficiency standards in the European Union? Analysis shows that this would both save money and reduce carbon dioxide emissions. A stock of high-efficiency homes and buildings (either built or retrofitted) would be attractive to owners and help kick-start the market from a stagnant position.

Energy and carbon efficiency improves national competitiveness; lowers prices for business, consumers and tourists; encourages foreign direct investment from new international employers; yields hundreds of millions of euro in energy savings; reduces the need for new infrastructure; helps achieve environmental targets and increases national security of supply. There are compelling reasons for individuals, businesses and government to make the right choices and implement bold energy efficiency measures with verve and confidence.

FIVE

Ireland's opportunity

'Trust yourself: every heart vibrates to that iron string. Accept the place that providence has found for you.'
– Ralph Waldo Emerson, *Self-Reliance*.

Ireland has an extraordinary opportunity to use its own natural renewable resources to achieve energy independence. The nation is endowed with winds that are among the strongest in the world. There is significant receipt of solar energy that can be captured by advanced solar technology despite the clouds. The waves that crash against the west of Ireland are some of the most powerful on the planet. Next-generation biomass grows prolifically and achieves among the highest yields in the world. In addition, geothermal energy and hydropower can also contribute to achieving energy independence.

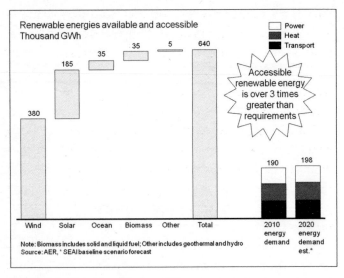

Potential contribution of renewable energy to Ireland's requirements

The available and practically accessible energy from primary renewable resources in Ireland is 675,000 GWh (or 675 billion kWh), based on summarised assessments in previous chapters. This enormous wealth of renewable energy is more than three times the amount of total energy required by the nation at present and in the foreseeable future.

The exceptional access to clean, indigenous energy provides Ireland with a unique opportunity not only to achieve energy independence but also to become a major exporter of energy to countries that are less well endowed. Ireland's accessible renewable energy is equivalent to an annual production of some 400 million barrels of oil, roughly the same as recent production in oil-rich Oman and Syria combined or just under half the production in Iraq or three quarters of production from the UK North Sea. Unlike oil, renewable sources will not dwindle as reservoirs are tapped but rather provide a consistent, safe and reliable source of energy for the duration of life on earth.

Future exports could provide transformative economic benefit to Ireland as international fossil fuel supplies inevitably fall and prices rise. Ireland's surplus accessible renewable energy, after total national energy requirements have been met, is roughly enough to supply the electricity needs of the UK and Belgium combined. The value of exports could be in the range of €30 billion to €50 billion per year depending on the future price of oil, which is in the region of 15 to 25 per cent of total Irish gross domestic product.

WHAT IS POSSIBLE IN THE NEAR TERM?

A full fifth of Ireland's total energy needs could be met by renewable energy sources by 2020. This would represent a practical and feasible first step towards energy independence.

Wind energy provides the greatest opportunity to reduce fossil fuel importation and to build clean electricity generation capacity. The potential for solar and wave energy is considerable but the development of cost competitive technologies is relatively slow. Therefore, just 3,000 GWh (or 3 billion kWh) of solar energy and a mere 1,000 GWh (or 1 billion kWh) of wave energy can be anticipated to be

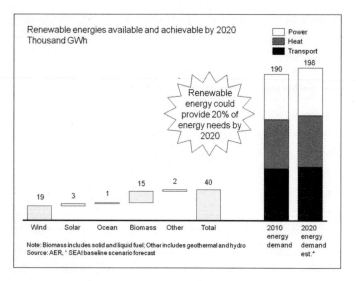

Renewable energies available and achievable by 2020
Thousand GWh

Power
Heat
Transport

Renewable energy could provide 20% of energy needs by 2020

190   198

19   3   1   15   2   40

Wind   Solar   Ocean   Biomass   Other   Total   2010 energy demand   2020 energy demand est.*

Note: Biomass includes solid and liquid fuel; Other includes geothermal and hydro
Source: AER, * SEAI baseline scenario forecast

Potential contribution of renewable energy to Ireland's
requirements by 2020

supplied by 2020. Solid biomass on the other hand can make a sizeable impact as it is readily available and cost competitive today. High-yield energy crops, wood and both municipal and agricultural waste provide a valuable opportunity. Liquid biofuel production in Ireland faces the challenge of inexpensive tropical ethanol imports. However, the advent of next-generation biofuels and a squeeze on international supply due to a rapid rise in global demand will provide near-term opportunities for sustainable biofuel production in Ireland.

Achieving a 20 per cent contribution to energy needs from renewable sources by 2020 would position Ireland to take a place among the highest ranked countries in the world for use of renewable energy.

VISION FOR ENERGY INDEPENDENCE

Some of the toughest challenges to achieving energy independence would be passed by meeting a 20 per cent renewable energy contribution by 2020. The hard work in building a core renewable energy infrastructure would be completed. More efficient planning and development processes would be in place. Companies and entrepreneurs would have confidence to build on the momentum of a growing market. The wells of renewable energy would have been successfully tapped and Ireland could continue to reap the abundant resources of wind, solar, wave and biomass that are available. A future vision of providing 80 per cent of Ireland's energy requirements from renewable resources can become a reality.

Growing renewable energy contributions to the 2050 targets shown in the illustration below will not mean a proportional increase in the number of wind turbines, wave devices or solar panels relative to those required to meet the 2020 target. Some increased deployment of equipment will be needed. However, significant contributions can be made by applying energy capacity and technology more efficiently and increasing overall productivity. Wind and wave energy can be stored as compressed air, hydrogen or water pumped to a higher elevation and there are multiple practical approaches available or being developed to accomplish this.

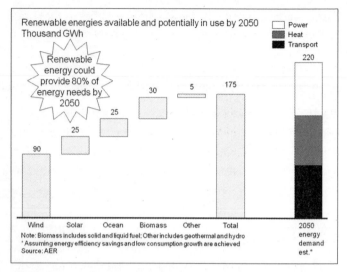

Vision for contribution of renewable energy and energy efficiency to Ireland's requirements by 2050

Increased penetration of electric vehicles can absorb renewable electricity and reduce dependency on imported oil. In addition, technology developments over the coming decades will enable the capture of more energy at less cost. Energy efficiency will play an important role in reaching energy independence. Efficiency measures that can save 20 per cent of energy consumption relative to 2010 have been identified and can be implemented, and energy consumption can be maintained at a stable level thereafter.

The vision for energy independence in Ireland is an aspiration. It is currently out of reach but within sight. There are many opportunities to say why it might not be possible. Leadership is the act of finding how to make it possible. In

1961, John F. Kennedy addressed a joint session of US Congress and declared his vision to 'put a man on the moon and return him safely by the end of the decade'. He did not know how this might be achieved. The technology at that point could not support it. Yet he promised that it would happen. By committing to the vision, he put in train all the activities that would be required to make it happen. And it did happen. Ireland can achieve energy independence if it commits to the vision and lives that commitment by continually working towards it.

ECONOMIC AND EMPLOYMENT OPPORTUNITY

The economic prize for working towards energy independence is great. The cost to achieve the first practical step of a 20 per cent renewable energy contribution and implement basic energy efficiency measures in Ireland by 2020 would be in the region of €15 billion, based on assessments for previous chapters. This is comprised of capital equipment, installation costs and ancillary services needed for wind, solar, wave, biomass and energy efficiency activities. The primary return on investment can be achieved from national and export sales of renewable energy. Additionally, the investment would result in an annual reduction of some €560 million of imported fossil fuels; an annual trading value of €180 million in carbon credits; annual receipt of €100 million in taxes; and an annual contribution to the economy of €340 million from efficient use of energy. The reduction in carbon dioxide emissions would result in an annual saving of the social cost of carbon of €760 million, as defined by the IPCC. These annual dividends would continue long after the period of intense capital investment ended.

A 2009 study by the High Level Group on Green Enterprise for the Irish government highlighted the potential to create more than 80,000 jobs in the coming years from activities in harnessing renewable energy (up to 50,000 jobs), retrofitting buildings for energy efficiency (up to 32,000 jobs) and managing waste in Ireland (over 1,500 jobs). In 2010 alone, 5,000 workers were employed in retrofitting 60,000 Irish homes and actually saving these homeowners and the state over €300 million in future energy costs – an outstanding 'win-win-win' outcome. The opportunity for these sectors to provide significant employment and grow the national economy is borne out in other countries.

Some 280,000 people were employed in Germany in renewable energy activities by the start of 2009, up from 249,000 a year earlier – in fact, an average additional 23,500 people have been employed in the renewable energy sector alone every single year in the five years to 2010. The sector had an operating turnover of €15.6 billion and investment of €13.1 billion in 2008. Significant national income was earned by exporting 40 per cent of produced solar technology and 70 per cent of produced wind technology. A further 1.8 million people are employed in broader environmental protection activities in Germany. Germany has also managed to reduce national carbon dioxide emissions by several hundred million tonnes over the five years to 2010. The Danish wind sector alone employs 28,000 people and contributes €5.7 billion to the economy. Japan aims to double employment in environmental and energy sectors to 2.8 million by 2020. South Korea has dedicated €27 billion to clean energy activities that will create almost a million jobs. In 2009, the US government committed $80 billion in funds and tax breaks with the potential to create 7.9 million jobs in green building and 2 million jobs in clean energy activities by 2015. In the same year, China committed €50 billion to ecological projects, Canada devoted $1 billion in green infrastructure and Italy set aside €2 billion for efficient cars and home appliances. In total, over €300 billion in international government funding has been committed in economic stimulus plans to clean energy and energy efficiency.

The ambitions of these countries create an international momentum that supports Ireland's own clean energy development. In particular, they constitute the world's fastest growing trading market into which Ireland can export technology and expertise. The global market for renewable energy and clean technology and services was worth about €1.2 trillion in 2009 and growth is estimated to continue at a rate in excess of 6 per cent every year. Ireland can tap into this huge global market by building on advantages such as the successful experience of technology-focused export businesses, a world class research and development base, a wealth and intensity of clean energy resources and an ability to promote Ireland's traditional 'green' brand.

CREATING AN ATTRACTIVE ENVIRONMENT FOR COMPANIES

Successfully achieving energy independence depends on the attractiveness of the energy opportunities to private sector competitors. Ireland operates in a free market economy and companies will only act on possibilities to create a positive economic return. Past experience has shown that state-owned or state-protected energy assets, especially those operating in near monopolistic conditions, may tend towards low levels of productivity, fail to compete in an international field and can become a money-sink for government. Furthermore, European Union competition rules constrain state aid that distorts market competition by providing unfair advantage to a select company or subset of companies. However, the government can provide open incentives for private companies to participate and thrive in clean energy development while maintaining a guiding regulatory hand to ensure maximum benefit to the state and its citizens. Irish governments and policymakers have already established several initiatives and these can be strengthened and further complemented. Suggested incentives and regulations (many already in place to be continued) include:

Incentives (the carrot):

(1) Continuation and increase of 'feed-in tariffs' that guarantee renewable energy supply prices and enable a positive return on investment.

(2) Faster, more efficient planning procedures for renewable energy projects, with fewer costs and delays.

③ Long-term, reduced interest rate loans for capital intensive renewable energy infrastructure facilitated by the Irish industrial development agencies, the National Treasury Management Agency and multilateral agencies such as the European Investment Bank.

④ Upgrade of the national grid to enable the addition of large-scale wind- and wave-generated electricity capacity.

⑤ Provision of a 'smart grid' capable of accepting local electricity production from sources such as micro-wind, farm biogas, combined heat and power and solar panels; Smart meters installed in every home and business, empowering and motivating consumers and businesses to manage energy efficiently and sell excess distributed electricity generation.

⑥ Faster grid connection process (e.g., current eighteen-month timeline for grid connection reduced to six months) and an improved regulatory model that further opens and modernises the electricity market.

⑦ Continued leadership in public sector procurement and asset management in supporting renewable energy and energy efficiency (the total Irish government pur - chasing budget is over €10 billion each year. The public sector is the largest landowner of property in the state; is also the largest tenant in the state and rents more offices and buildings in the state than any other entity; and also owns the largest fleet of transport vehicles in the state. These positions provide govern- ment and the wider public sector with significant leverage to drive renewable energy and energy efficiency markets).

(8) Lower vehicle registration taxes for flexible biofuel and electric cars.

(9) Continued grants for energy efficiency improvements (e.g., SEAI 'greener homes scheme').

(10) Continued 'accelerated capital allowance scheme' enabling investment in equipment that improves energy efficiency to be offset against corporate taxes.

(11) Free audits of commercial premises to identify energy savings and efficiency opportunities.

(12) Employment incentives for new renewable energy and energy efficiency roles.

(13) Government and industry co-funding of research and development in renewable energy and energy efficiency. Continued funding tied strictly to commercialisation metrics (e.g., patent applications, technology licensing and spin-out company revenues).

Regulations (the stick):

(1) Continued application of tax on carbon use, with levy income redirected to renewable energy and energy efficiency activities.

(2) Additional 'carbon reduction levy' for largest energy users (e.g., greater than 5 GWh per year) and carbon dioxide emitters (e.g., greater than 40,000 tonnes per year) that is redistributed to those companies demonstrating the greatest energy and carbon dioxide annual reductions, published each year in a league table.

(3) Rules to ensure that new buildings meet the highest efficiency standards in Europe and old buildings demonstrate annual progress in increasing efficiency with penalties for non-compliance.

(4) Stipulation that all new building developments use a set minimum amount of renewable energy (e.g., 25 per cent).

(5) Obligation on electricity suppliers to produce a minimum of 50 per cent electricity from renewable sources (an increase from current government target of 40 per cent).

(6) Enforcement of national waste management policy that further drives activities to 'reduce, reuse and recycle' resources, including water.

The cost of incentives provided can be paid back by way of levies raised through the regulatory obligations, as well as through increased tax receipts, higher employment and reduced costs in waste management, pollution, health and international carbon penalties.

CHALLENGE MUST BE MET BY A NATIONAL MOBILISATION

The challenge of achieving a 20 per cent overall renewable energy contribution and the associated economic and employment benefits, let alone full energy independence, cannot be underestimated. It has been shown that Ireland has sufficient renewable energy resources to meet the target several times over. The challenge, therefore, is one of mobilisation to capitalise on the opportunity, taking the starting position into account.

In 2008, the Irish government signed an EU Climate Change package along with all other EU governments. The EU targets in this package require Ireland to fulfil at least 16 per cent of its energy consumption using renewable sources and to reduce its 2005 level of greenhouse gas emissions by 21 per cent by 2020. In the year that Ireland committed to these targets, it fulfilled just 4.5 per cent of its energy consumption using renewable sources; its contribution of renewable sources to gross electricity consumption was only 11.9 per cent; and its greenhouse gas emissions were 6.7 per cent above 2005 levels.

The task to surpass the EU targets and demonstrate global leadership by achieving a 20 per cent renewable energy contribution given the starting position is therefore a major challenge. It requires significant investment and the collaboration of industry, government and communities over many years – no less than a national mobilisation of the human and natural resources that would be required to defend against serious external threats. In this case, the threats include the risks of an international energy crisis; the shutdown of energy supplies to an island at the end of the pipelines and shipping lanes; and destructive climate change.

The prize for successful mobilisation – national economic transformation, major sources of employment, freedom from dependence on imported energy, averted threats to security, positive investment payback, significant efficiency savings and global clean energy leadership – is unparalleled.

MOBILISING SUCCESSFULLY IN THE PAST

Ireland has risen to major challenges to its identity, independence and economic survival in the past. In facing these challenges, the mobilisation of people and resources on a national level has led to some of Ireland's greatest achievements.

The wildfire growth of national identity during the Gaelic Revival at the turn of the twentieth century; the energy independence achieved from building what was then the world's largest hydroelectric plant in the 1920s; the energy and resource independence achieved during the National Emergency in the 1940s; and the transformational programmes for national recovery and economic and social development in the 1980s and 1990s in response to crushing debt and unemployment all reflect a spirit of collective action to build a better nation and contribute to a better world. It is this spirit and successful mobilisation of resources and people on a national level that is needed again to seize Ireland's greatest opportunity – energy independence using clean energy.

Gaelic Revival

Three inspiring movements built on indigenous cultural resources and converged to forge a new spirit of identity in Ireland at the turn of the last century. The first movement was initially named the 'Gaelic Athletic Association for the Preservation and Cultivation of National Pastimes'. It was founded in 1884 by Michael Cusack, a schoolteacher from

County Clare, and Pat Nally, from a Mayo farming family, and began to have a major impact on Irish culture from 1900. Cusack and Nally shared a passion to promote Gaelic sports and moreover to provide a forum for participation for every person in every community. The GAA took root in every corner of the country and became a focus for a renewed and positive national identity. In Cusack's words, the GAA 'swept the country like a prairie fire'. One commentator noted that 'the country was soon humming with interest and activity'. Today, the GAA is the largest amateur sporting body in the world and an international model for community-based collective action.

The second movement, the Gaelic League, was founded in 1893 by Douglas Hyde, a writer from County Roscommon who was later to become Ireland's first President (1937–45), and Eoin MacNeill, an Irish scholar from County Antrim. Hyde and MacNeill wished to preserve Irish as the national language and promote the study and advancement of Irish literature. More than 600 branches of the league were established throughout the country and 50,000 active members were enlisted by 1904.

The third movement was known as the Irish Literary Revival. The founders, including W. B. Yeats, Maud Gonne, Douglas Hyde and Lady Gregory, described in their charter that they were determined 'to build up a national tradition, a national literature, which shall be nonetheless Irish in spirit for being English in language'. The Literary Revival produced a powerful body of avant-garde writing that shaped an important strand in international literature for decades and established Ireland as a world leader in culture and arts development. The players of the Abbey Theatre, established in 1904, were also global ambassadors of Irish theatre and set the international performance bar on stages throughout Britain and the United States.

Ireland has proven that it can be a world leader in cultural development, sports, literature, theatre and music, based on the wealth of its creative resources. It can translate this collective action to set the international bar for clean energy development, based on the wealth of its natural energy resources.

The Shannon Scheme

Two years after the Civil War and four years after the War of Independence, Ireland took a position of world leadership to establish energy independence. In 1925, the Irish Free State embarked on one of the largest engineering and construction projects in the world at the time to harness the clean energy of the River Shannon so that electricity could be brought to the entire country.

Only a third of Dublin houses had electricity installed at the time and much of rural Ireland had no electricity infrastructure at all. Access to a reliable electricity grid would bring light and energy, which would improve the quality of life and facilitate industrial development in every corner of the new state.

The idea to build the world's largest hydroelectricity plant on the Shannon near Limerick had been presented to the government by Drogheda-born engineer Thomas McLaughlin. He shepherded the concept through a rigorous six-month feasibility assessment with Siemens-Schuckert, a German-based international engineering firm. A contract was subsequently awarded to Siemens-Schuckert following an international tender. The cost would be IR£5.5 million and one third of this was earmarked for expanding the national grid. This cost amounted to roughly 20 per cent of the government's revenue budget in 1925, underlining both the project's significance and risk to the nascent state. The government faced fierce criticism from many who labelled the project a white elephant and a waste of scarce funds as the country emerged from economic and social depression. However, the leaders of the day passionately believed in the benefits that would accrue from energy independence and pressed on. *Financial Times* journalist W. M. Harland wrote of the leaders in 1928 that 'they seized on this chance of showing what they can do ... never forgetting the practical benefits ... for agricultural and industrial development. They have had thrown on their shoulders the not easy task of breaking what in reality is an enormous inferiority complex and the Shannon Scheme is one and probably the most vital of their methods of doing it.'

Over three and a half years, government and industry partnered to mobilise a workforce of up to 5,200 workers, 1,770 railway wagons, 130 steam locomotives, 31 barges and multiple tools and other equipment. Workers shifted 8.8 million cubic metres of earth and rock, dug 14 km of canal to divert 90 per cent of the River Shannon across a 34 m drop, laid 96 km of dedicated railway, extended the electricity grid into remotest Ireland and constructed the world's largest hydroelectric power plant. The Electricity Supply Board (ESB) was established in 1927 to oversee the Shannon Scheme and general national electricity operations. The Scheme earned an international reputation due to its scale and ambition and over 250,000 visitors joined tours of the site during construction.

The plant opened in October 1929 and provided up to 96 per cent of the electricity required by the nation. It was hailed as a phenomenal engineering, commercial and political success throughout the world. Franklin D. Roosevelt wrote to the ESB seeking information on 'the magnificent Shannon Scheme' and the Electricity Act, 1927. Herbert Hoover had taken an interest in the project when Thomas McLaughlin visited the US in the mid-1920s. The size of the Shannon plant was surpassed only by the building of the Hoover Dam, the most expensive US federal project completed in the twentieth century.

Seventy-five years after the opening of the Shannon Scheme, the ESB and Siemens were awarded the Milestone and Landmark Awards by the Institute of Electrical and Electronic Engineers and the American Society of Civil Engineers. These awards were in recognition of the Shannon Scheme's role as a 'model for large-scale electrification projects worldwide' and its 'immediate impact on the social, economic and industrial development of Ireland.' The true significance of the Shannon Scheme as a contribution to society and international development is reflected in the fact that Milestone Awards had previously been given for the invention of colour television, the space shuttle and Japan's bullet train while Landmark Awards winners included the Eiffel Tower, the Golden Gate Bridge and the Panama Canal.

Newly independent Ireland had successfully established itself as an international leader for development and energy self-reliance. The nation's leaders had recognised the immense potential energy of the one of the country's best natural resources, faced down the critics and seized the opportunity with determination and self-belief. The Shannon plant still provides clean electricity today, though at less than 2 per cent of national requirements due to the increase in demand in the intervening years. Almost a century later, the same spirit of 'what is possible' is needed to harness the new streams of clean energy that can once more set Ireland on a surge of social, economic and industrial development.

National Emergency

The outbreak of the Second World War as Britain and France declared war on expansionist Germany led to the announcement of a National Emergency at 5 a.m. on Sunday 3 September 1939. The Emergency signalled the need for the nation's politicians, industry and citizens to mobilise together and defend against the greatest threat to the young state's security and self-sufficiency.

Invasion by either Germany or Britain appeared to be imminent on many occasions during the course of the war. Hitler himself stated that 'the occupation of Ireland might lead to the end of the war. Investigations are to be made.' A possible invasion of Ireland as a back door to Britain, codenamed 'Operation Green' was planned by the German military after the Luftwaffe failed to smash the RAF in the Battle of Britain. The operation outlined an invasion by six German columns and contained minute details and aerial photos of Irish airfields, harbours, transport routes and Irish defence force bases. At the same time, a British contingency plan to invade Ireland from the north, overrun the country, seize the capital and the three strategic ports at Cobh, Berehaven and Lough Swilly were relayed to the government by Irish contacts in the British forces.

The threat was most destructively manifest in the dropping of German bombs on Wexford in August 1940, killing three girls; on Drogheda, Wexford, Wicklow, Kildare, Dublin and Carlow, killing three people in the first two days of 1941; and on Dublin in May 1944, killing thirty-four people and injuring ninety. The German government expressed regrets for the Dublin bombings, 'which may have been due to high wind'.

While the threat from rampant German forces was frightening, damaging economic pressure also came from neighbouring Britain. Ireland had prohibited vessels of war, submarines and aircraft of any belligerent in the war to use Irish territorial waters and airspace in order to uphold Ireland's position of neutrality in compliance with international law. This incensed British Prime Minister Winston Churchill who wanted access to Ireland's strategic ports to support the defence of his nation. He decided to use economic pressure to force Ireland to abandon its position of neutrality and yield use of the ports to the British navy.

Churchill's government first imposed sanctions by restricting the licensing of ships bound for Ireland, chartered through the UK Shipping Office. This drastically reduced Ireland's ability to import necessary goods from any location. Churchill subsequently ordered the restriction of certain British exports to Ireland, including food, equipment and industrial goods, on which Ireland was highly dependent. Britain cut petrol supply to Ireland in half and coal supply was limited to a bare minimum of lowest quality grade. By the end of 1941, petrol pumps had run dry and motorists were stranded throughout the country. Homes went cold through the winter as coal was no longer available for heating. The national train service ground to a halt. City trams and Dublin electricity generation, partially dependent on imported coal, faltered. White bread disappeared entirely from shelves and tea was rationed to 0.5 ounces per week. Ireland's economy and way of life were buckling with the stress of resource starvation under an effective blockade. Further pressure resulted from an outbreak of foot-and-mouth disease and the consequent slaughter of 19,000 cattle and 5,000 sheep.

The national leaders, industry and citizens pulled together in the face of this severe crisis. The leaders of the three main political parties, De Valera, Cosgrave and Norton, held a cross-party defence conference and stood together on public platforms calling for collective action and volunteers. Regional commissioners were appointed and empowered to ensure efficient resource management. Over 30,000 volunteers enlisted in the defence forces during the Emergency and 98,000 joined part-time defence forces. Tens of thousands of citizens mobilised and invested time and energy in drilling and manoeuvres that simulated war scenarios.

Ireland turned to its indigenous sources of fuel to keep industry and national transport moving. Turf was cut and burned instead of imported British coal. Troops were deployed to harvest bog lands to fuel trains and electricity plants. People were encouraged to cut turf in the Dublin Mountains for their own domestic heating and the Phoenix Park was lined with stacks of drying sods that ran for miles.

The nation also turned to its indigenous agricultural resources to mitigate the loss of imported food and grain. Vast tracts of land were converted from grazing to tillage and the area under wheat almost tripled from 90,000 hectares in 1938 to 260,000 acres in 1944.

The government invested in several key resources to bolster self-sufficiency. It supported textile and leather manufacturing to provide clothing during the Emergency years; invested in new ships and established Irish Shipping Limited to build the national commercial fleet and increase trading with the US, Africa and Portugal; commissioned Irish naval vessels for the first time; and set up the Scientific Research Bureau to develop new technologies that would contribute to national self-sufficiency.

Above all, Ireland maintained a steadfast commitment to neutrality and independence in democratic deference to overwhelming public opinion. Energy and resource independence was achieved in the face of consistent and crippling economic sanctions. The success of this independence further strengthened a relatively new Irish identity and self-belief. It continues to shape national character seventy years later. Such independence and self-reliance on indigenous resources must again be applied in the face of the impending energy threat.

Economic transformation

Ireland's economic transformation in the late 1980s and 1990s has been a model to nations throughout the world. By the mid-1980s, the economy was on the verge of collapse. National debt as a proportion of gross national product (GNP) had increased from 67 per cent in 1973 to 105 per cent in 1983. Taxation as a proportion of GNP increased from 31 per cent in 1973 to 42 per cent in 1985. Unemployment rose to over 16 per cent and the country was experiencing mass emigration of many of its best minds.

The deep crisis forced national leaders to set aside sectional interests. The government, employers and trade unions agreed a Programme for National Recovery in a spirit of unified purpose. Industrial development policy continued to foster a high-tech, knowledge-based, export-driven economy. Employment incentives provided for foreign direct investment, especially focused on key areas of high international demand such as the pharmaceuticals and information and communication technology sectors.

Corporation tax on trading income was set at 12.5 per cent in the mid-1990s and extended to the services sector, consolidating the advantages of a 10 per cent manufacturing tax that had been set in 1981 and had encouraged almost every major pharmaceutical, medical devices, computer and software multinational company to establish facilities in Ireland. Over IR£4 billion (€5 billion) was invested in research and development during the 1990s.

The government increased investment in education to build Ireland's human resource skills and provide the knowledge and competitive advantage to create successful enterprises. The proportion of the population receiving third level education doubled between 1980 and 2000 to more than 23 per cent, making Ireland second only to Belgium in the European Union for the proportion of young adults with third level degrees. Six out of ten third level degrees were in science, engineering or business studies and increasing numbers of students mastered a high level of proficiency in international languages. IMD scored Ireland higher than its European neighbours and the United States on the degree to which its educational system met the needs of a competitive economy. These strong educational standards and labour force flexibility were major incentives for Irish and international companies to establish and grow from an Irish base.

These actions and investments led to a period of exceptional economic growth during the 1990s. The average annual rate of growth of gross domestic product (GDP) was 9 per cent, outstripping the European average by a factor of three. Unemployment tumbled from 16 per cent in 1993 to less than 4 per cent in 2001, effectively achieving full employment. Over the same period, the debt to GNP ratio fell from 90 per cent to less than 40 per cent.

The focus on core resources and competitive advantages in Ireland – namely: a highly educated workforce; an increasing spirit of enterprise; independent fiscal policy; high-tech, export-oriented industrial policy; and access to the European Union – enabled Ireland to transform itself from an insular, low-productivity economy with crippling debt and unemployment to a high-growth, low-debt, full-employment economy and one of the most open and competitive trading nations in the world within a decade.

Ireland became a case study for governments, strategists, business professors and students throughout the world. Delegations from multiple international governments and countless organisations came to Ireland with the sole purpose of studying the Irish economic success and learning how to apply the lessons in their respective countries. The country's success story hit the covers of the world's best-selling business and economics magazines. Ireland held its head high among the nations of the world.

The demise of the 'Celtic Tiger' came later. The primary driver for the reversal of fortunes was an external global credit crisis, simmering for a number of years and boiling over from 2008. Three key internal flaws added to the fall in growth rates from 2007 had little connection with earlier success factors. These were: an advantageous tax rate for property development, which contributed to exuberant price rises and an ultimate property bubble, that was introduced in 2001;

massive personal and commercial leveraged debt that created a severe risk in the event of capital markets contraction, which were mainly accumulated post 2000; and a 'growth at all costs' attitude and lax regulatory enforcement in the banking sector that led to over-lending, questionable transactions and huge resultant losses, which was prevalent primarily during the property boom from 2000.

The painful recession was shared by almost every developed country. It cannot detract from the extraordinary success of Ireland's economic transformation in the previous decade. The long-term basis for a healthy economy was set in place during that period. The foundations of education, open and competitive trading and enterprise spirit remain firm. Ireland will learn from the recession and thrive again in the years ahead. Growth will be more sustainable. A major impetus for renewed development can be the harnessing of clean energy from Irelands own natural resources.

BELIEVE

Ireland has demonstrated the confidence and ability to assume a role of world leadership in these examples and on many other occasions. In each case, Ireland has successfully applied indigenous resources – creativity, hydropower, agricultural produce, land and people. Ireland is blessed with another set of resources that is waiting to be tapped – wind, wave, sun and biomass.

Past successes have often occurred when the nation's back was against the wall. Ireland may thrive as the underdog with everything to play for in order to survive. The corollary may apply, i.e., that confidence and collective action are not in abundance without a sense of emergency or crisis. The reticence may partly be a legacy of historical imprint – the psychology of oppression – a reluctance to cut dependence on a source of sustenance, i.e., imported energy, and take advantage of prolific indigenous resources. Clean energy is a chance for economic and energy freedom.

If there is one word to describe what Ireland must do to realise its greatest opportunity, it is 'believe'. *Believe* in the wealth of natural resources sufficient to gain energy independence. *Believe* in the ability and determination of people to build and shape the world's first clean energy economy. *Believe* that the example Ireland can set will become the model for all nations that wish to gain independence and avert a global energy and climate crisis. *Believe* that these things are not only possible but that they lead the way on a path to greater equity, health and peace throughout the world.

Remaining Questions

I reland faces its greatest challenge and greatest opportunity in how it will address its future energy needs. Will
it succumb to the path of least resistance and continue to
suckle from the sumps of foreign oil? Will it embrace the
wealth and diversity of indigenous renewable resources,
investing above the cost of oil now for long-term independence and economic returns? The answer will define the
nation for the next century and beyond. Its character will be
set as a follower or a leader.

There are deeper, greater questions posed by the energy
dilemma of increasing energy needs in the face of dwindling
fossil fuel supply and climate change. They face all people
and all nations and Ireland's actions will shape and influence
how other people and nations will think about them:

➡ When and how do individuals and nations decide that
they have gathered, stored and consumed 'enough'?
➡ How can the quintessential human urge to gratify
needs and desires in the present be balanced to ensure

that future standards of living and welfare are not compromised?

➡ How can justice and protection be provided to those constituents of life such as nature and living species other than humans, which do not have a voice in human systems of governance such as the electoral platform or the boardroom?

➡ Will the distribution of energy resources be done with a spirit of equity, according to needs, or instead under a system of resource colonialism, managed by those with the greatest financial and military power?

➡ Should the positive growth of a nation be defined almost entirely in economic terms (e.g., growth in GDP) or should aspects of sustainability and quality of life be added to the definition?

➡ How can the current democratic system, shaped by a consistent *short-term* focus on re-election, be developed and reconciled with the *long-term* vision, decisions and investments that are required to solve the energy dilemma?

➡ How can nations achieve energy independence in a way that does not invoke protectionism and compromise the shared benefits of international comparative advantage and trading?

➡ Will the strongest human instinct, survival, translate from the individual, who may not feel an imminent threat to her or his daily life because consequences of an energy crunch and climate change are not immediately apparent, to the collective, which is under threat at a global level and is required to act with urgency?

These questions and others that will shape human society are thrust upon this generation as a result of the end of the cheapest and most abundant source of energy that mankind has managed to harness in history. They will be answered within a lifetime. The answers will emerge either as a consequence of passive human behaviour, guided by unthinking instinct, or by the active agency of thoughtful response and collective action. The latter requires national leaders to take a stand and set example. Leaders with vision, confidence and natural resources are needed. Ireland can be such a leader if it chooses.

How these questions are answered will not only solve the energy dilemma for better or worse but will shape humanity. Energy is the first of the key resources to come under threat on a global level. The same tensions will emerge on a global level (having emerged to date only on local level) for water, food, land and air as they become internationally constrained. They will be constrained not just to one player in a system, but to all players. How energy is managed will set the first precedent for how all future global crunches will be managed. Solving energy issues will shape the nature of all future existence.

Notes

ENERGY UNITS

The derived unit for energy under the international system of units (SI) is the joule (J).

Energy can also be denoted variously in watt-hours (1 Wh = 3,600 J), British thermal units (1 Btu = 1,055 J), foot-pound force (1 ft lb = 1.356 J), horsepower-hours (1 horsepower-hour = 2,684,519 J), barrels of oil equivalent (1 boe =~ 6.2 x 10^9 J) and calories (1 thermochemical calorie = 4.184 J).

The rate of energy conversion is known as power or energy capacity and its SI derived unit is the watt (W).

Because energy is in a constant state of flux, it is described both in terms of the capacity for conversion (using the unit W) and the actual energy converted over a period of time (using the unit Wh).

This book uses both primary units, Wh and W, when describing energy. However, for the sake of consistency in comparing across energy forms, the units are always related to amounts that are commonly used. For instance, 'kWh' is commonly used in recording energy converted over time in a household electricity bill (e.g., a 1,000 W heater turned on for 1 hour uses 1 kWh of electricity), while 'MW' is commonly used in describing the size of a power plants (e.g., the Moneypoint power plant in County Clare has a capacity of 915 MW).

Energy generally occurs in large amounts and the notations are shown below:

1 Wh = Wh (watt-hour)
10^3 Wh = kWh (kilowatt-hour)
10^6 Wh = MWh (megawatt-hour)
10^9 Wh = GWh (gigawatt-hour)
10^{12} Wh = TWh (terawatt-hour)
10^{15} Wh = PWh (petawatt-hour)
10^{18} Wh = EWh (exawatt-hour)
10^{21} Wh = ZWh (zettawatt-hour)
10^{24} Wh = YWh (yottawatt-hour)

OTHER SOURCES OR CARRIERS OF RENEWABLE ENERGY

Geothermal

The heat that emanates from the earth's core can be captured in the form of direct heating for buildings and steam-driven electricity generation. Geothermal energy was used for hot water and cooking by the ancient Romans and has been used by the native peoples of North America, Iceland and New Zealand for thousands of years. Today it is used in over seventy countries for commercial energy generation but provides only 0.2 per cent of total global electricity capacity. Annual growth in capacity is slow at 2.5 per cent because there is a limited number of suitable locations. The operating cost of geothermal-generated electricity is in a similar range to onshore wind-generated electricity and the costs for heating are minimal in places that have naturally rising geothermal heat, such as Iceland, where it is the source for more than 50 per cent of heating requirements.

Geothermal energy provides a consistent supply twenty-four hours a day with no carbon dioxide emissions. On a local level, every home can install a geothermal pump for domestic heating and increase energy efficiency relative to oil or gas. The

International Energy Agency (IEA) estimates that less than 5 per cent of the world's surface is suitable for very large scale geothermal energy production. Therefore, this source is likely to remain a niche, albeit useful, contributor to large-scale energy needs.

In Ireland, no major sources of concentrated geothermal energy have yet been harnessed. However, a subsurface fault line running below Newcastle to Blackrock in County Dublin has been identified as a potential source of geothermal energy for district heating. Boreholes have been drilled to a depth of 1,400 m, where hot water was tapped and awaits further development. Low-level geothermal heating can be supplied from ground source heat pumps for all homes and buildings. The Sustainable Energy Authority of Ireland provides support to those wishing to install heat pumps.

Hydro

Energy can be harnessed from water falling from rivers or dams as it passes through turbines to generate electricity or, as it has been used for millennia, to do mechanical work turning mill wheels. It is estimated by the World Energy Council that 16,000 TWh (16,000 billion kWh) of hydro energy is available per year, using current technology. This is slightly less than the world's current electricity consumption and about 12 per cent of total energy needs. Hydro energy already provides a significant contribution to world energy, generating some 15 per cent of global electricity. However, most of the world's largest and most suitable reservoirs are now in use. Up-front capital costs for hydroelectricity plants are often high and deter investors but lifetime operating costs can be less than all other types of electricity generation.

In Ireland, hydroelectricity plants on the River Shannon, Turlough Hill in County Wicklow and several smaller locations provide between 1 and 2 per cent of Ireland's electricity needs each year. Remaining natural sites are small in scale and unlikely to prove economical, despite the fact that up to 200 river locations were once used for mills that provided energy for the grain, cloth and brewing industries. Several western coastal valleys carved by retreating glaciers have been identified by the 'Spirit of Ireland' project leaders as possible locations for creating large reservoirs. Water could be pumped into these valleys using excess wind capacity. Such reservoirs would help smooth the supply of energy from intermittent winds and could provide a large contribution to national energy needs as well as potential exports.

Hydrogen

A hydrogen fuel cell works by combining hydrogen and oxygen in an electrochemical process to create electricity. This principal has been applied by astronauts to create electricity in space since the 1960s. The use of hydrogen as an energy carrier remains far from broad commercial feasibility but it enjoys a growing interest from oil companies and car manufacturers. For instance, in 2007, BMW released a 7-series model than runs on a hydrogen-powered fuel cell, proving that the fuel and technology can be applied successfully.

The cost of producing hydrogen fell by 40 per cent in the decade to 2010 as technology improved and the fuel is cost competitive with gasoline. However, the cost of the fuel cell itself in cars is about ten times the cost of internal combustion engines. There are clear potential advantages to hydrogen fuel cells if the overall costs can be reduced. For instance, the only by-products of operation are water and heat. Operation is quiet and efficient. Hydrogen production could be used as a means to smooth the intermittent nature of wind and solar energy supply. Excess wind or solar energy could be applied to extract hydrogen from water by way of electrolysis. The energy stored in hydrogen could later be released when wind or sunshine are in short supply.

Unfortunately, the production of hydrogen itself is energy intensive. If sources other than water are used to produce hydrogen, such as natural gas or coal, carbon dioxide is released and the overall environmental benefits are undermined. In addition, creating a new distribution and storage infrastructure for hydrogen that would rival the existing infrastructure for fossil transport fuels is an expensive task. Despite the challenges, it is likely that hydrogen will feature as a complementary clean energy technology in the future.

Bibliography and Sources

NOTE: Estimates given in this book of global and Irish available and accessible energy resources reflect a balanced assessment of multiple published and peer-reviewed sources, which often provide varying data. The estimates have been further tested by first principles calculations and reference to existing industry field experience.

1. ENERGY IS . . .

Kierkegaard, Soren, *Fear and Trembling* (Copenhagen, 1843)

Russell, Bertrand, *History of Western Philosophy* (London, 1945)

Vitruvius Pollio, Marcus, *De Architectura* (Rome, *c*.27); Translation, Granger, F. *Vitruvius* (London, 1933)

Ovidius Naso, Publius (Ovid), *Metamorphoses* (Rome, AD 8)

Serway and Faughn, *College Physics* (Philadelphia, 1999)

Cambell and Reece, *Biology* (San Francisco, 2005)

Lewis, Finlay R., ed., *Focus on Non Verbal Communication Research* (Louvain, 2007)

Baker, Joanne, *Fifty Physics Ideas* (London, 2007)

Rose, David, *Learning About Energy* (New York, 1986)

McMullan, J. T., Morgan, R., Murray, R. B., *Energy Resources* (London, 1983)

McGown and Bockris, *How to Obtain Abundant Clean Energy* (New York, 1980)

Ubbelohde, A. R., *Man and Energy* (London, 1963)

Messel, H., ed., *Energy for Survival* (Sydney, 1979)

Lenihan, John and Fletcher, William, eds., *Environment and Man, Volume One* (Glasgow, 1975)

2. FOSSIL FUEL'S FATAL FLAWS

Getty, J. Paul, *My Life and Fortunes* (London, 1964)

International Monetary Fund, *World Economic Outlook Database* (Washington D.C., 2009)

BP, *BP Statistical review of world energy June 2009* (London, 2009)

Roberts, Paul, 'Tapped Out', *National Geographic* (Washington D.C., 2008)

Yergin, Daniel, *The Prize: The Epic Quest for Oil, Money and Power* (New York, 1991)

American Association of Petroleum Geologists, *Giant Oil and Gas Fields of the Decade 1990–1999* (Kansas City, 2003)

US Central Intelligence Agency (CIA) *World Factbook 2009*, www.cia.gov

Van der Veer, Joroen, Shell Chief Executive, 'Easy oil has probably passed its peak', *Financial Times* (London, 2006)

De Margarie, Christophe, Total Chief Executive, 'Optimistic case for maximum oil output is 100 million barrels', quoted in *National Geographic* (Washington D.C., 2008)

Mulva, James, ConocoPhilips Chief Executive, 'Output will stall at 100 million barrels a day, where is all that going to come from?', *National Geographic* (Washington D.C., 2008)

National Geographic, www.ngm.nationalgeographic.com

Diamond, Jared, *New York Times*, *What's your Consumption Factor* (New York, 2008)

Jiabao, Wen, *2009 Summer Davos speech*, www.chinaview.cn (DaLian, 2009)

WRAP, Food Waste data UK, www.wrap.org.uk

Dewan, Shaila, 'Hundreds of coal Ash Dumps Lack Regulation', *New York Times* (New York, 2009)

Appenzeller, Tim, 'The High Cost of Cheap Coal', *National Geographic* (Washington D.C., 2006)

Klare, Michael, *Blood and Oil* (London, 2004)

Riegle, Donald, D'Amato Alfonse, United States Senate, US Committee on Banking, Housing and Urban Affair with respect to Export Administration, *US Chemical and Biological Warfare Related Dual Use Exports to Iraq and their Possible Impact on the Health Consequences of the Gulf War* (Washington D.C., 1994)

Douthwaite, Richard, *The Growth Illusion* (Dublin, 2000)

Meadows, Donella, Meadows, Dennis, Randers, Jorgen, Behrens III, William, *The Limits to Growth* (London, 1972)

Meadows, Donella, Meadows, Dennis, Randers, Jorgen, *Limits to Growth, the 30 year update* (London, 2005)

Burroughs, William James, *Climate Change, A Multidisciplinary Approach* (Cambridge, 2007)

Archer, Daniel, *Global Warming, Understanding the Forecast* (Chicago, 2007)

Strzepek, Kenneth and Smith, Joel, *As Climate Changes* (Cambridge 1995)

United Nations Environmental Panel, World Meteorological Organisation, *Intergovernmental Panel on Climate Change (IPCC) Climate Change 2007: Fourth Assessment Report of 2007* (Geneva, 2007), www.ipcc.ch

Carbon Dioxide Information Analysis Center, http://cdiac. ornl.gov

Maslin, Mark, *Global Warming, A Very Short Introduction* (Oxford, 2004)

Peixoto and Oort, *Physics of Climate Change* (New York, 1992)

3. IRELAND'S VORACIOUS THIRST FOR ENERGY

Sustainable Energy Authority of Ireland, www.seai.ie

Sustainable Energy Ireland, *Energy in Ireland 1990–2008; Energy Statistics 1990–2008; Energy Forecasts for Ireland to 2020* (Cork and Dublin, 2009)

Sustainable Energy Ireland, *Energy in Ireland 1990–2007* (Cork and Dublin, 2008)

Sustainable Energy Ireland, *Energy in Ireland Key Statistics 2007* (Cork and Dublin, 2008)

Travers, John T., Address to United Nations Development Programme Conference, *Ireland's Economic Development* (Washington D.C., 2000)

Travers, John, *Driving the Tiger, Irish Enterprise Spirit* (Dublin, 2001)

Forfás, Amárach Consulting, *A Baseline Assessment of Ireland's Oil Dependence* (Dublin, 2006)

Bord Gáis, www.bordgais.ie

Bord na Móna, www.bnm.ie

Eurostat, www.eurostat.ec.europa.eu

ConocoPhillips, www.conocophillips.com

Institute of International and European Affairs, *The Climate Change Challenge Strategic Issues, Options and Implications for Ireland* (Dublin, 2008)

Environmental Protection Agency, prepared by the National University of Ireland, Maynooth, *Climate Change Scenarios and Impacts for Ireland* (Wexford, 2003)

Environmental Protection Agency, prepared by the National University of Ireland, Maynooth, *Climate Change Refining the Impact for Ireland* (Wexford, 2008)

The Irish Academy of Engineering, *Critical Infrastructure Adaptation for Climate Change* (Dublin, 2009)

4. DRINKING WATER FROM A FIRE–HOSE: ALTERNATIVE ENERGY OPTIONS

Wind

O'Hare, Greg and Sweeney, John, *The Atmospheric System* (Essex, 1995)

Simmons, Daniel, *Wind Power* (New Jersey, 1975)

Cheremisinoff, Nicholas, *Fundamentals of Wind Energy* (Ann Arbour, MI, 1979)

McWilliams, Brendan, *Weather Eye* (Dublin, 1994)

Lu, Xi, McElroy, Michael and Kiviluoma, Juha, Harvard University, 'Global Potential for Wind Generated Electricity', *Proceedings of the National Academy of Sciences* (Washington D.C., 2009)

World Energy Council, *Survey of Energy Resources 2007*, www.worldenergy.org

Carr, Peter, *The Big Wind* (Belfast, 1991)

Sustainable Energy Ireland, Howley, Martin, Ó'Gallachóir, Brian, Dennehy, Emer, *Energy in Ireland 1990–2007* (Cork, 2008)

Irish Wind Energy Association, www.iwea.com

Sustainable Energy Authority of Ireland, www.seai.ie

Sustainable Energy Ireland, ESB International, UCD Energy Research Group, *Renewable Energy Resources in Ireland for 2010 and 2020* (Dublin, 2004)

ESB International and ETSU, *Total Renewable Energy Resource in Ireland* (Dublin, 1997)

Renewable Energy Strategy Group (Fitzgerald, John and Ó'Gallachóir, Brian), *Strategy for Intensifying Wind Energy Deployment, Renewable Energy Strategy* (Dublin, 2000)

House of the Oireachtas, Joint Committee in Climate Change and Energy Security, third report, *Meeting Ireland's Electricity Needs Post 2020* (Dublin, 2009)

Danish Energy Agency, Energy Statistics, www.ens.dk

Center for Politiske Studier (CEPOS), *Wind Energy, The Case of Denmark* (Copenhagen, 2009)

Sun

McVeigh, J. C., *Sun Power* (Oxford, 1983)

Sharp, www.sharp–world.com/solar

Moore, Patrick, *Suns, Myths and Men* (London, 1968)

Koupelis, Theo and Kuhn, Karl, *In Quest of the Universe* (Boston, 2007)

King, Henry, *Pictorial Guide to the Stars* (New York, 1967)

Gribbin, John, *The Strangest Star* (London, 1980)

World Energy Council, *Survey of Energy Resources 2007*, www.worldenergy.org

International Energy Agency, Swens, Job, *PV in The Netherlands 2006* (Utrecht, 2007)

European Solar Thermal Industry Federation, *Solar Thermal Markets in Europe, Trends and Market Statistics, 2008* (Brussels, 2009)

Greenpeace, European Solar Thermal Power Industry Association, Solar Paces, *Concentrated Solar Power Now* (Amsterdam, 2005)

European Photovoltaic Industry Association, *Global Market Outlook for Photovoltaics until 2013* (Brussels, 2009)

Newton, James, *Uncommon Friends, Life with Thomas Edison, Henry Ford, Harvey Firestone, Alexis Carrel and Charles Lindbergh* (Orlando, 1987)

Sustainable Energy Authority of Ireland, www.seai.ie

Sustainable Energy Ireland, ESB International, UCD, *Renewable Energy Resources in Ireland for 2010 and 2020* (Dublin, 2004)

Ocean

Brooke, John, *Wave Energy Conversion* (Amsterdam, 2003)

World Energy Council, *Survey of Energy Resources 2007*, www.worldenergy.org

Kampion, Drew, *The Book of Waves* (Boulder Colorado, 1997)

LeBlond, P. H. and Mysak, L. A., *Waves in the Ocean* (Amsterdam, 1978)

The Pembina Institute, www.pembina.org

US Department of Energy, www.energysavers.gov

Thorpe, T. W., *An Overview of Wave Energy Technologies* (London, 1999)

MacKay, David, *Sustainable Energy Without the Hot Air* (Cambridge, 2009)

Fitzgerald, Richard and Slater, Kelly, Element Pictures, *Waveriders*, Lighthouse Cinema Dublin and www. waveridersthefilm.com

Irish shipwrecks database, www.irishwrecksonline.net, accessed July and August 2009

Sustainable Energy Authority of Ireland, www.seai.ie

Sustainable Energy Ireland, ESB International, UCD, *Renewable Energy Resources in Ireland for 2010 and 2020* (Dublin, 2004)

ESB International and ETSU, *Total Renewable Energy Resource in Ireland* (Dublin, 1997)

Department of Communications, Marine and Natural Resources, Marine Institute, Sustainable Energy Ireland, *Ocean Energy in Ireland* (Dublin, 2005)

Sustainable Energy Ireland, *Tidal and Current Energy Resources in Ireland* (Dublin, 2004)

Tedlow, Richard S., *Giants of Enterprise, Seven Business Innovators and the Empires They Built* (New York, 2001)

Biomass

Harlan, Jack, *Crops and Man* (Madison, WI, 1992)

Khoshnakht, Korous and Hammer, Karl, *Genetic Resources and Crop Evolution, How Many Plant Species Are Cultivated?* (Dordrecht, 2008)

Vaclav Smil, *Biomass Energies* (New York, 1983)

Slesser, Malcolm and Lewis, Chris, *Biological Energy Resources* (London, 1979)

Reynolds, S. G. *et al*, Food and Agriculture Organisation (FAO), *Grassland of the World* (Rome, 2000)

US Department of Energy, *Carbon Dioxide Emissions from the Generation of Electric Power* (Washington D.C., 2000)

UK Department of Environment, Food and Rural Affairs, *Biomass Taskforce Report to Government* (London, 2005)

World Energy Council, *Survey of Energy Resources 2007*, www.worldenergy.org

Coford, Knaggs, Gordon and O'Driscoll, Eoin, *An Overview of the Irish Wood Based Biomass Sector* (Dublin, 2008)

Klvac, Radomir, Ward, Shane, Owende, Philip M. O. and Lyons, John, 'Energy Audit of Wood Harvesting Systems', *Scandinavian Journal of Forest Research* (London, 2003)

Department of Communications, Marine and Natural Resources, *Bioenergy Action Plan for Ireland* (Dublin, 2006)

Department of Environment, Food and Rural Affairs, UK, *2009 Guidelines to DEFRA/DECC's GHG Conversion Factors* (London, 2009)

Sustainable Energy Ireland, *Liquid Biofuels Strategy for Ireland* (Dublin, 2004)

Travers, John, AER Limited, Cork County Council, *Mallow Sugar Factory, Ethanol Production Evaluation Study* (Dublin and Cork, 2006)

Department of Communications, Marine and Natural Resources, *Bioenergy in Ireland* (Dublin, 2004)

Gorecki, Paul, Lyons, Sean, Acheson, Jean, Economic and Social Research Institute, *An Economic Approach to Municipal Waste Management Policy in Ireland* (Dublin, 2010)

Nuclear

Murray, Raymond, *Nuclear Energy* (New York, 1993)

OECD, *Nuclear Energy in Perspective* (Paris, 1989)

Krauskopf, Konrad, *Radioactive Waste Disposal and Geology* (London, 1991)

BBC, *Horizon: Windscale, Britain's Biggest Nuclear Disaster* (originally broadcast on 8 October 2007)

World Nuclear Association, www.world-nuclear.org, *Tokaimura Criticality Accident* (London, 2010)

Moneypoint Generating Station, *2008 Annual Environmental Report* (Dublin, 2009)

Poolbeg Generating Station, *2008 Annual Environmental Report* (Dublin, 2009)

Environmental Protection Agency, www.epa.ie

Nuclear Energy Agency OECD, *Addressing Climate Change* (Paris, 2009)

O'Flaherty, Tom, Institute of Engineers in Ireland, *Nuclear Energy: Is it an option for Ireland?* (Dublin, 2009)

Irish Business and Economics Confederation (IBEC), www.ibec.ie

Barre, Betrand, *Pros and Cons of Nuclear Power* (Dublin, 2007)

Comby, Bruno, *Will Nuclear Have a Role in Ireland?* (Dublin, 2007)

Lovelock, James, *The Vanishing Face of Gaia: A Final Warning* (New York, 2009)

Energy and carbon efficiency

Sustainable Energy Ireland, *Demand Side Management in Ireland, Evaluating Energy Efficiency Opportunities* (Dublin, 2008)

Department of Communications, Energy and Natural Resources, *Maximising Ireland's Energy Efficiency, The National Energy Efficiency Plan 2009–2020* (Dublin, 2009)

Sustainable Energy Ireland, McKinsey *Ireland's Low Carbon Opportunity* (Dublin, 2009)

The Sustainable Energy Authority of Ireland, Power of One, www.powerofone.ie

Carbon Trust UK, www.carbontrust.co.uk

Walsh, Conor, Jakeman, Phil, Moles, Richard, O'Regan, Bernadette, *A Comparison of Carbon Dioxide Emissions Associated with Motorized Transport Modes and Cycling in Ireland* (Limerick, 2008)

Paddy Comyn, 'Surprise Carbon Offenders', *The Irish Times* (Dublin, 2009)

European Union green paper on energy efficiency, *Doing More with Less* (Brussels, 2006)

Brown, Lester, *Plan B 4.0, Mobilising to Save Civilisation* (New York, 2009)

Curtin, Joseph, Institute of International and European Affairs (IIEA), *Jobs, Growth and Reduced Energy Costs: greenprint for a national energy efficiency retrofit programme* (Dublin, 2009)

5. IRELAND'S OPPORTUNITY

Ritch, Emma, Cleanteach Group, *Germany, US, Australia Inject Stimulus Spending into Cleantech*, www.cleantechn. com

Federal Ministry for the Environment, Nature Conservation and Nuclear Safety, *Development of Renewable Energies in Germany in 2008* (Berlin, 2009)

US Green Building Council, *Green Jobs Study* (Washington D.C., 2009)

Report of the High Level Group on Green Enterprise (chair, Joe Harford), *Developing the Green Economy in Ireland* (Dublin, 2009)

Travers, John, *Driving the Tiger, Irish Enterprise Spirit* (Dublin, 2001)

Hammons, Tom, Institute of Electrical and Electronic Engineers Power Engineering Review, *Shannon Scheme for the Electrification of the Irish Free State* (London, 2002)

Siemens, www.siemens.com

Harland, W. M., *Financial Times*, referenced in Manning, Maurice, McDowell, Moore, *Electricity Supply in Ireland, The History of the ESB* (Dublin, 1984)

Carroll, Joseph, *Ireland in the War Years* (Newton Abbott, Devon, 1975)

Fisk, Robert, *In Time of War, Ireland, Ulster and the Price of Neutrality 1939–45* (London, 1983)

Index